WHAT ARE THE ODDS?

THE REAL RULES OF CHANCE THAT GOVERN OUR LIVES

Tim Glynne-Jones

CHARTWELL
BOOKS, INC.

This edition printed in 2011 by
CHARTWELL BOOKS, INC.
A Division of **BOOK SALES, INC.**
276 Fifth Avenue Suite 206
New York, New York 10001 USA

Copyright © 2011 Arcturus Publishing Limited
26/27 Bickels Yard, 151–153 Bermondsey Street,
London SE1 3HA

ISBN-13: 978-0-7858-2803-7
ISBN-10: 0-7858-2803-6
AD001651EN

Printed in China

CONTENTS

INTRODUCTION

Whether or not you are a risk-taker by nature, you will no doubt pay some attention to the odds in just about everything you do. You may call it probability, likelihood, chance or risk, but whether you're planning to swim across shark-infested Sydney Harbour or you're just wondering if you should leave the house without an umbrella, you will weigh up the odds.

We have the Vikings to thank for the word 'odds'. The singular 'odd' is believed to stem from Old Norse, meaning triangle. Forms of the word were used to describe triangles of all sorts, from landforms to spearheads. Triangles are three-sided and three-cornered – they are the embodiment of the number three, and so the word odd soon became synonymous with three. The Vikings liked threes because threes could get things done. When two people disagreed on a matter, the third – the odd man – would have the casting vote. The odd man soon applied to any uneven number, and thus the word assumed the meaning we give it today.

The plural, 'odds', came to mean the difference between two things that were not equal. So in betting today, any bet where the return matches your stake is known as evens; anything else is known as odds.

'What are my chances of getting eaten by a shark?' 'How likely is it to rain today?' These are the sorts of questions that preoccupy the minds of men and women throughout the world, and have done throughout history. Stone Age man would have peered out of his cave in the morning and wondered what his chances were of catching a mammoth for tea. And at the same time he might have allowed his mind to touch on the likelihood of becoming dinner for a sabre-toothed tiger.

This book does not seek to provide you with reliable odds for every eventuality in life. Instead, it looks at the world and life and all the fantastic, the phenomenal, the tragic, the bizarre and even the mundane occurrences that leave us scratching our heads and asking ourselves what the odds of that must have been.

British Prime Minister Benjamin Disraeli contended that statistics were just a form of lies. He was right in so far as, no matter how thoroughly researched they are, statistics can be misleading. Just because, statistically speaking, a tossed coin has a **one in two** chance of coming up heads, doesn't mean a tail will be followed by a head every time. That random element, the exception that proves the rule, is what gives the odds their excitement. The chance that a 100-1 outsider might come home first is what keeps life from being boringly predictable.

Statistical data can provide a rough idea of the probability of certain events taking place, and no more. At other times you just have to take a more philosophical view and ask yourself, 'Am I feeling lucky?'

Tim Glynne-Jones

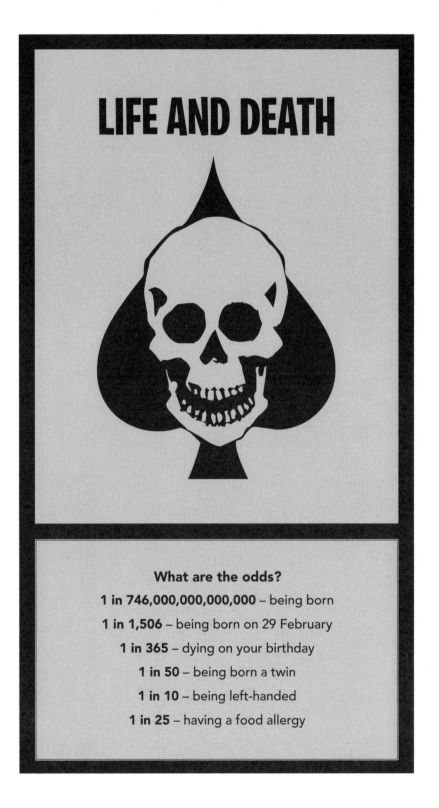

LIFE AND DEATH

What are the odds?

1 in 746,000,000,000,000 – being born

1 in 1,506 – being born on 29 February

1 in 365 – dying on your birthday

1 in 50 – being born a twin

1 in 10 – being left-handed

1 in 25 – having a food allergy

THE MIRACLE OF BIRTH

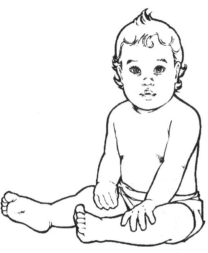

Let's begin at the beginning. As you know, you are the result of one of your father's sperm fertilizing one of your mother's eggs. Your mother had all her eggs when she was born, so you could have been any one of the 400 or so eggs that she produced in her fertile life. In other words, the egg that made you had roughly a **1 in 400** chance.

But sperm, now that's another matter. The average man is reckoned to produce about 3,500,000,000,000 sperm in his lifetime.

That's approximately 150 million per day between the ages of 11 and 75. However, as we are talking about the chances of you being born, we have to look at the potential reproductive period of your father and mother together. The average age of fathers is 32, and if we take 16 as the minimum age, we can assume 48 to be the maximum, which happens to be around the average age for the menopause.

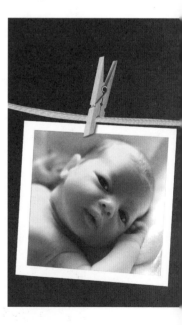

Therefore, of the 64 years in which the average man produces sperm, the potential for fatherhood spans an average 32. So we can discard half of those 3.5 trillion sperm, giving us odds for the sperm that made you of **1 in 1.75** trillion.

Multiply those by the odds of your sperm fertilizing your egg:

1/1,750,000,000 x 1/400 = 1/700,000,000,000,000

But this is just the probability of your conception. Recent figures from the National Vital Statistics System in the USA show that 93.8 per cent of foetuses survive through to birth, further reducing your chances of being born to:

1/700,000,000,000,000 x 93.8/100

= 93.8/70,000,000,000,000,000

= 1 in 746,000,000,000,000

Of course, these odds assume the existence of your father and mother in the first place. When you take into account the odds against them having been born, and the odds of their parents, and their parents' parents, and the odds of them getting together with each other, out of all the people on Earth, you realize that your birth defied odds that make the likelihood of your existence practically 0.

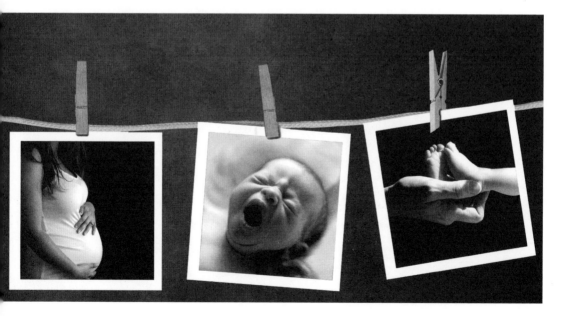

WHAT ARE YOU LIKE?

Being born is just the start of the complex probability maelstrom that governs who you are. Next up is the major decision (more major in some societies than others) as to what gender you happen to be.

Across the world, the average ratio of new born boys to girls remains fairly constant at about **1.05 to 1**. Put another way, for every 1,000 girls born, there are 1,005 boys born.

Why this should be is uncertain. It could be nature's way of balancing up the overall population, since male life expectancy is lower than female. According to UN population statistics, the global divide between males and females of all ages is more or less the reverse of the difference at birth. That is, for every 1,000 men in the world there are 1,004 women.

But this discrepancy fluctuates a great deal from country to country. The countries of the former Soviet Union have a disproportionate ratio of women to men. In Russia, Latvia and Estonia, for example, women outnumber men by **1.17 to 1**. If it's nature's plan to couple males and

females in order to propagate the species, this means that for every 100 women with a male partner there will be 17 without.

Meanwhile, in China, India and the Arabic states the opposite is true. High volumes of imported male workforce in the UAE, Bahrain, Qatar etc have created a huge gender imbalance in the population. Qatar is the most extreme case, with a male:female ratio of **3 to 1**.

TWO OF A KIND

Twins play an effective part in evening up the gender balance. In 50 per cent of cases twins will be one of each sex.

The chances of being born a twin are **1 in 50**.

The chances of being an identical twin are **1 in 285**.

That's on a global scale, but there are anomalies. The Yaruba people of West Africa have a twin birth rate of **1 in 22**.

First World countries have experienced a significant rise in their twin population due to two factors. The increased use of fertility treatment is one factor. Practices such as IVF carry a higher than average chance of producing twins because more than one fertilized egg is placed in the uterus in order to increase the chance of a successful conception.

The other factor is the age of the mother. A study in the UK found that twins made up just 0.63 per cent of births by mothers aged below 20, while the figure leapt to 21.7 per cent in mothers aged between 35 and 39.

This factor is further illustrated by the fact that 62 per cent of twins and 70 per cent of triplets, but only 48 per cent of singletons, are born to mothers aged over 30.

The odds of being one of triplets are just below **1 in 4,000**.

And the odds of being a quadruplet are less than **1 in 785,000**.

YOU NEED HANDS

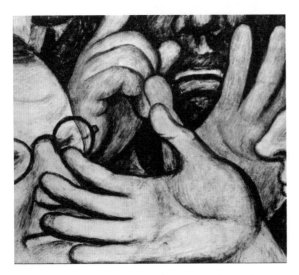

Left-handers make up 10 per cent of the world's population, a proportion that remains fairly constant throughout the world.

But how are your chances of being left-handed affected by your parents? Is a left-handed parent more likely to have a left-handed child? Can two left-handed parents ever have a right-handed child?

The answer to that last question is a definite yes; in fact, they usually do. A mere two per cent of left-handers come from two left-handed parents. The vast majority – 75 per cent – are born to parents who are both right-handed.

However, parents who are both left-handed do have a statistically greater chance of having a left-handed child: about **1 in 5**, compared to the average of **1 in 10**. There are just far fewer of them. And parents who are both right-handed have a slightly below average chance of spawning a lefty: about **9 in 100**.

Where one parent is left-handed and the other right-handed, the odds of having a left-handed child are right on the mean: **1 in 10**.

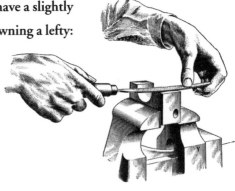

BLUE GENE BABY

Parents are often surprised by the colour of their children's eyes. When two brown-eyed parents produce a blue-eyed baby, the first thought may be to blame the milkman. After all, isn't it a well-known fact that the brown-eyed gene dominates all other colours?

This may be true, and there are certainly more brown-eyed people in the world than any others, but the odds of a brown-eyed couple giving birth to a blue-eyed baby are actually quite good.

Babies are often born with blue eyes because the pigment in the iris isn't activated until it is exposed to light, but that's not our concern here. It's a question of genetics. To keep things simple, let's concentrate on the two main genes that govern eye colour. One can be either brown or blue, the other can be green or blue. Everybody inherits one of each type from each parent, so we all have two of each type.

Where a green and a blue are together, the green will dominate. Similarly, a brown will dominate both green and blue. Therefore, in order to have blue eyes you must have a blue version of all four genes.

How is it possible to inherit this from brown-eyed parents?

In order to have brown eyes, a person only needs one brown gene as that will dominate any others. So the parents in this case could have one brown and one blue of gene type A, and one green and one blue of gene type B, for example. That will be enough to give them both brown eyes, but, as this diagram shows, the child still has a **1 in 16** chance of inheriting the two blue genes from each parent.

Each pair of letters represents the genes that have been inherited by each person. The child then has 16 different genetic combinations that he or she could inherit, one of which is four blue genes.

Parent A: **Bb+Gb**   Parent B: **Bb+Gb**

Child: **bb**   +   **bb**

[B=Brown, b=blue, G=green]

From this you can also get an idea of how an eye colour can 'hide' within a family for generations, until a child inherits the necessary combination of genes for that colour of eyes. Because brown is the dominant gene, it can be passed down through generations in company with a blue gene, producing brown eyes all the way, until one day a child inherits the blue gene rather than the brown and combines it with another blue or a green, and hey presto, blue or green eyes, just like the milkman!

NB: What I've described is a very simplistic view of the genetics of eye pigmentation. In fact, there are further determining factors, including other genes, and instances where a gene may be turned off, or mutated.

WHERE DID THAT RED HAIR COME FROM?

Red hair is associated with Celtic blood, and when a blond and a brunette give birth to a redhead, confusion reigns.

Hair colour is even more complex than eye colour. In fact, they haven't yet pinpointed exactly how hair colour is determined, except in the case of red hair. Like eyes, hair colour is caused by the pigment melanin. Blond and black hair is coloured by eumelanin. Red hair is coloured by pheomelanin. We all have both pigments in differing degrees. Blonds only have a small amount of both. Black-haired people have a lot of eumelanin. Redheads have little eumelanin and a lot of pheomelanin.

The reason most people don't have red hair is because of a gene, MC1R, which allows pheomelanin to be converted to eumelanin. In some cases this gene is mutated. People with freckles are often found to have one mutated MC1R gene. However, having just one is not enough to make you a redhead. It takes two mutated MC1R genes to stop the conversion of pheomelanin to eumelanin, and that's when red hair occurs.

Therefore, parents who do not have red hair but give birth to a red-haired child must have one mutated MC1R gene and one working one each, and pass on their mutated genes to the child. The odds, in this case, are **1 in 4**.

THE BALD FACTS

There is a school of thought that says any colour of hair is better than no hair at all, so people, especially men, are often keen to know their chances of going bald.

Male pattern balding, or androgenic alopecia, to give it its scientific name, is most prevalent in Caucasian men, but it affects all races to a fairly similar degree.

In Caucasians, the odds of losing your locks increase in direct proportion with age. That is, 30 per cent of men in their 30s, 50 per cent of men in their 50s and 60 per cent of men in their 60s experience hair loss.

Male pattern balding has been traced to the androgen dihydrotestosterone (DHT), often referred to as the male sex hormone. This is the basis of the theory that bald men are more virile. DHT has a dual effect of stimulating hair growth as well as exfoliating, which explains why it's quite common to see men with more hair on their chest than they have on their head. Bald men are often hirsuit in other areas.

WHAT ARE THE ODDS OF TOPPING 6FT?

The height we grow to is determined by more than just genetics. True, a tall mother and father will produce tall offspring, but factors like nutrition and environment also play a part. Tallness is undoubtedly the result of both nature and nuture.

In the Netherlands, for example, the average height for a man is 6ft (183cm). Yet a century ago, a quarter of Dutchmen measured less than 5ft 2in (157cm). What precipitated this phenomenal growth spurt was a vast improvement in diet.

In fact, we should probably look at it the other way round: for a large part of human history, poor diet has kept the average height artificially low. The Dutch are statistically the tallest nation on Earth and that applies to the women too, measuring an average 5ft 7in (170cm). That's 4in (10cm) taller than the average man in Bolivia – the shortest country on Earth and, paradoxically, the highest. The Bolivian city of Potosi stands at 13,420ft (4,090m) above sea level, 790ft (240m) higher than the highest town in the Himalayas. But its men average 5ft 3in (160cm) tall, and its women just 4ft 8in (142cm). Perhaps altitude plays a part – the air is thinner up there, and the Netherlands also happens to be one of the lowest countries on Earth (hence the name).

IS IT SAFE FOR COUSINS TO HAVE CHILDREN?

If it's not safe for siblings to have children together, why is it okay for cousins? Surely there's a similar risk.

In fact, the risk of a child inheriting a genetic disease from parents who are cousins is tiny compared with the risk when the parents are siblings. If the risk among unrelated couples is about 3–4 per cent, then the risk among cousins is reckoned to be 5–7 per cent (roughly double).

With siblings, the risk can be as high as 25 per cent if one of their parents is a carrier.

In most cases, in order for a child to be born with a genetic disease, it has to inherit the same bad gene from both its mother and father, because bad genes are usually recessive – in other words a good gene will override a bad one, which constantly improves the odds of inheritance.

Let's suppose the father of a boy and girl carries a bad gene but their mother doesn't. His children each have a **1 in 2** chance of inheriting that gene. The odds of both of them inheriting it are **1 in 4** (multiply the odds).

They now have one bad version of the gene and one good one each, just like their father. If they then grow up and have a child together, that child has a **1 in 4** chance (25 per cent) of inheriting two bad genes and, therefore, having the disease.

Now, if we apply the same scenario to cousins, and the brother and sister each has a child with unrelated partners, each one has a **1 in 2** chance of passing on their bad gene to their child, and the odds that both pass on their bad gene to both children (the cousins) are **1 in 4**.

Finally, those cousins have a child. The odds that it will inherit a bad gene from one parent are **1 in 2**. The odds that it will inherit a bad gene from both parents are **1 in 4**.

So we have three generations of **1 in 4** chances, which we multiply together to give **1 in 64**.

Instead of 25 per cent in the case of siblings, we're looking at 1.56 per cent.

Of course, not all cases of genetic disease follow this inheritance pattern. Another carrier could come into the equation at any point. But this illustrates quite clearly that the risk facing first cousins, though double the risk for unrelated couples, is a fraction of the risk facing siblings.

DEMONIC POSSESSION OR OCD?

Obsessive Compulsive Disorder (OCD) is one of those conditions that seem to have arrived with us in the last few decades. The numbers of people diagnosed with OCD has increased exponentially over the last 50 years, with 3 per cent of the world's population now estimated to have it. But this rise is down to the heightened awareness of the condition and a greater openness about it.

OCD manifests itself in the form of fixations and repeat behaviour. It was once believed to be a symptom of demonic possession and was treated by exorcism. Now that it is better understood, we can look back through history and identify numerous famous figures who probably suffered from OCD: Samuel Johnson, Michelangelo, Beethoven, Einstein, Howard Hughes… right up to David Beckham and countless Hollywood stars.

There is evidence that OCD is genetic: 20–35 per cent of children with OCD are related to someone else with the condition. However, it doesn't appear to be 100 per cent genetic. Studies of identical twins have found that if one has it, the other has an 80–87 per cent chance of having it. Identical twins share the same DNA, so it if was entirely genetic, both would have it in every case. However, the fact that it is in part genetic is shown by the findings that in non-identical twins, dual incidence occurs in only about **1 in 2** cases.

DON'T PLAY WITH YOUR FOOD

The odds of having a food allergy are on the rise. Whether the cause of this increase is genetic or environmental has yet to be determined. So far it looks like being both. **1 in 25** people in the First World are now reckoned to suffer with a food allergy, but among children under 3 that rate is up at **1 in 17**.

Each year, about **1 in 10,000** people in the First World are hospitalized by an anaphylactic reaction to some food they've eaten. In fact, they don't even have to eat it. Simply by coming into contact with a food they're allergic to, an anaphylaxis sufferer can develop a potentially fatal allergic reaction. About 0.5 per cent of cases are currently fatal.

Nutritional hazards

The food most associated with anaphylaxis is the peanut according to the Food Allergy & Anaphylaxis Network (FAAN). The others are milk, egg, tree nuts, fish, shellfish, soy and wheat. An allergic reaction takes place when the body's immune system overreacts to the presence of a substance it takes to be hostile. Peanut allergy affects nearly **1 in 100** people and only **1 in 5** are expected to outgrow it in adulthood. Research is still trying to identify potential hereditary traits, but statistics show that having a close relative with the condition increases your chances of being born with it by a factor of 14. Environment is also a factor, though. Mothers who eat peanuts during pregnancy increase their child's allergy risk fourfold, and breastfeeding mothers who eat peanuts double it.

SORRY, SON, NO BIRTHDAY THIS YEAR

You have to feel sorry for people who are born on February 29 in a leap year. They only get to celebrate their actual birthday every four years. So why do we have leap years?

A leap year is a mathematical adjustment that was worked into the Gregorian calendar to account for the fact that a year is not a nice round 365 days in length, but 365 days, 5 hours, 48 minutes and 47 seconds. If we carried on counting each year as 365 days, the whole thing would slip out of synch over time.

So because a year is approximately 365 days and a quarter, adding an extra day every four years keeps things fairly neatly on track. Further fine-tuning is provided by the omission of the leap year every 100 years, and its inclusion every 400 years. So, while 2100 won't be a leap year, 2000 was.

If we concern ourselves only with the present day, with people born in the last 100 years and people who will be born in the next 50, the odds of being born on 29 February are simple to calculate. All we have to do is count the number of days in four years: **(365 x 3) + 366 = 1,461.** This gives us odds of **1 in 1,461.**

However, if we disregard the present day and look at the probability in general, we have to take into account the fact that there are only 97 leap years in 400.

So we calculate the number of days in 400 years thus:

97 x 366 + 303 x 365 = 146097

Then we take the number of 29 Februarys in those 400 years = 97. And that gives us our probability: **97/146097 = 1/1506.1546.** Probability theory is great for calculating these sorts of odds in principle, but when we look at factual evidence a different picture emerges. With the above calculations we assumed that there is an equal probability of being born on any day of the year, but this isn't actually the case.

In fact, fewer people are born in February than any other month. A recent selection of UN statistics (see graph, right) shows the results for births in the USA, Great Britain, Japan, Australia and Russia.

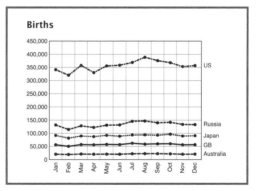

While the most popular months varied between July (UK) and October (Japan), February was the most unpopular month in every case.

While the theoretical average of being born in any given month is 8.33 per cent (**1 in 12**), the average of being born in February in these five countries was 7.44 per cent (or **1 in 13.44**).

Which begs the question, why do fewer people make babies in the merry month of May? Are they trying to protect their offspring from the disappointment of a 29 February birthday, or are they too busy enjoying the sunshine?

HAPPY BIRTHDAY TO YOU TOO

Do you find it remarkable how many people you know who share a birthday? In fact, if you know 23 people, the chances are 50-50 that at least two of them will share a birthday. Put another way, among an average soccer World Cup squad (23 players), the odds of two players sharing a birthday are **1 in 2**.

Given that there are 365 days in a year (ignoring leap years), you would expect it to be more unlikely than likely that two people in a World Cup squad of 23 should share a birthday. In fact, it appears highly unlikely, since the chance of Player A sharing a birthday with any of the other 22 is 22/365, which is less than **1 in 16**.

However, we're not just focusing on Player A. We're looking for a matching pair from any of the 23 players present. So we need to look at the number of possible matches. We calculate that by multiplying 23 (the total number) by 22 (the number of possible matches for each person) and dividing by 2 (since Player A + Player B is the same as Player B + Player A), which gives us 253.

So now that we know there are 253 possible matches and 365 days in the year, the odds start to look rather better.

Now we can begin to calculate the odds by starting small and getting bigger. We have to assume that there is an equal probability of being born on any day of the year. First, we are going to turn our problem round so that we're calculating the probability that no two players in the squad share a birthday.

Starting with just two players, the chance of them not sharing a birthday is 365/365 x 364/365. In other words, the first person can be born on any day of the year, leaving 364 days if the second person's birthday is not to coincide. This gives us a 364/365 chance of the two

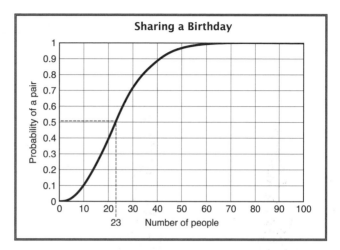

not sharing a birthday. And since the chances of something happening and not happening amount to 1, we can deduce that the chance of them sharing a birthday is **1 in 365**.

This method of calculation comes into its own as the number of people increases. If the group contained five people, the odds of them not sharing a birthday would be easy enough to calculate: 365/365 x 364/365 x 363/365 x 362/365 x 361/365. And so it goes on, with the last sum in the list always being 365-n+1/365, where n is the number of people in the group.

Feel free to do the sums yourself, but at the top of this page there's a graph that shows the results for you.

As you can see, with a group of 23 players the probability of two of them sharing a birthday is just over **1 in 2** (50.7%). And although you need a group of 366 to achieve a probability of 1 (100%), any group larger than 57 people has more than a 99 per cent chance of a shared birthday.

I'M A BELIEVER

In spite of the development of civilization, the advancement of technology, space exploration and other giant leaps, the likelihood of growing up as an adherent to a religion is increasing. According to 'World Christian Encyclopedia: A comparative survey of churches and religions – AD 30 to 2200', only 14 per cent of the world's population in the year 2000 claimed not to follow an organized religion, and this number was falling.

Most people adhere to the religion they were born into and currently the most common religions are Christianity, Islam and Hinduism, in that order. The survey found that Christians account for about a third of the world's population, Muslims about a fifth and Hindus an eighth. However, these proportions are changing rapidly, with the birth rate among Muslims exceeding that among Christians, to the extent that the percentage of Muslims in the world is expected to overtake that of Christians within the next 20 to 30 years.

So while the odds of being a Christian

Top 10 Religions of the World

1. Christianity	**1 in 3**	7. New religions	**1 in 50**
2. Islam	**1 in 4**	8. Sikhism	**1 in 277**
3. Hinduism	**1 in 8**	9. Judaism	**1 in 455**
4. Chinese folk religion	**1 in 17**	10. Spiritism	**1 in 524**
5. Buddhism	**1 in 17**		
6. Tribal religions	**1 in 25**	[source: www.religioustolerance.org]	

currently stand at about **1 in 3**, and the odds of being a Muslim at **1 in 4**, these figures will have reversed by the year 2030.

While a third of the world's population declare themselves to be Christian, a lot fewer actually go to church on a regular basis. In most countries where Christianity is the predominant religion, fewer than **1 in 2** people claim to go to church at least once a week. In the United States, for example, you have a 44 per cent chance of being a regular churchgoer, and even in Italy, the home of Roman Catholicism, that figure is only one per cent higher.

Nigeria is the leading nation for church attendance, with 89 out of every 100 people attending church once a week or more. Ireland too has a high percentage, with 84, but other than the Philippines, South Africa, Poland and Puerto Rico, every other country sees less than 50 per cent attendance at Christian churches.

EVERYBODY HAPPY?

Are you born happy? Or is happiness something you achieve? There are countless factors that affect happiness, but there is one surefire way of gauging whether people are happy: ask them.

The World Values Survey did just that, and discovered that the nation with the most happy people was Iceland. In Iceland you have a 94 per cent chance of being happy. In fact, northern Europe had a very bright outlook indeed. Sweden, Denmark and the Netherlands all returned figures of 91 per cent. The Aussies are generally happy, according to 9 out of 10 respondents. In the more developed countries, you've got at least a 75 per cent chance of being happy.

However, not so many of them would go so far as to say they were very happy. Only 42 per cent of Icelanders put themselves in that bracket. For extreme happiness, Venezuela's the place to be. There, 55 per cent of people declare themselves to be very happy. At the opposite end of the happiness scale, Eastern Europe takes the mouldy biscuit. A brooding 62 per cent of Bulgarians claim to be not very happy, with the next 13 countries in the unhappy list all being in Eastern Europe. India is the first to buck the trend, with only 30 per cent claiming unhappiness.

SATISFACTION

Where you live has a significant bearing on happiness. The average person living in the First World rates their level of satisfaction with their life at around **7.5 out of 10**.

In Eastern Europe and Africa, the average satisfaction rating is around **4 out of 10**.

FREEDOM

Being free to choose your own path in life is another influence on happiness. In the Land of the Free, Americans rate their actual freedom to make their own decisions at **7.6 out of 10**. Canada is the same, but the country with the greatest sense of freedom is Finland, at **7.7 out of 10**.

NATIONAL PRIDE

In the USA and Ireland, national pride abounds. The Americans and Irish have a 77 per cent chance of feeling proud of their nationality. However, it seems that national pride isn't a necessity in feeling satisfied or happy. In many of those Northern European countries where happiness is rife, only around 45 per cent of people are proud of where they come from.

That said, ask them if they would fight for their country and 80–90 per cent say they would.

SPACE

Overcrowding can be a major cause of unhappiness, especially if you're having to share your house with too many people. In Sweden and Denmark, only **1 in 20** households consist of five or more people, and the average domestic situation is two rooms for each person. It's a similar story throughout the developed world, and in stark contrast to Pakistan and India where on average three people share a room.

Sometimes you just need to get away from it all. In Australia and Iceland, you stand a good chance, with fewer than three people to every square kilometre (0.4 sq mile). But if you're happy in the hustle and bustle, go to Monaco. You'll be sharing your square kilometre with 17,240 other people.

SPACED

Some people try to boost their sense of happiness by taking drugs. So how likely are you to join them? Again, it depends where you live. In New Zealand, for example, more than **1 in 5** people are, or have been, cannabis users. Australians are almost as keen, with 18 per cent of the population having smoked cannabis.

In the USA, the chances of becoming a cannabis smoker are **1 in 8**. But in northern Europe, they don't need the weed to keep smiling. Fewer than **1 in 25** Scandinavians ever smoke dope.

DRUG ABUSE

According to UN estimates, 4.3 per cent of the world's adult population (over 15) consumes illicit drugs.

Cannabis is by far the most common substance, accounting for 69 per cent of all drug abuse.

1 in 470 adults use heroin.

1 in 320 use cocaine

1 in 130 use amphetamines

Drug abuse is far more prevalent in men: 4 out of 5 deaths from drug abuse are males. And the problem is most severe in North America, Europe and Australasia.

1 in 5 students in Europe have used cannabis at least once.

In Russia, the number of drug users has doubled to more than 5 million in the last 10 years. That's 3.5 per cent. Half of them are addicts.

1 in every 1,750 Russians are registered as drug addicts annually.

1 in 5 of Russia's drug addicts are school children.

Russia is now the world's most heroin-addicted country, consuming one fifth of the total heroin production each year.

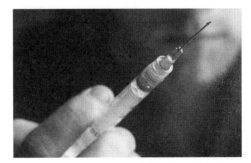

In the USA, the odds are **1 in 66** that you will try heroin at least once in your lifetime.

If you go to jail, there's a **1 in 5** chance that you will use heroin.

HAPPY HOLIDAYS

Going on holiday is supposed to be the highlight of the year: a chance to forget about your worries and your cares, soak up a bit of sun, eat some good food, meet some nice people and generally boost your sense of wellbeing. But there are so many things to worry about on holiday.

UP, UP AND AWAY

The flight is the main cause for concern, and it's not just the thought of crashing that brings on the travel anxiety. Will the flight be on time?

On average, 80 per cent of flights depart and arrive on time.

That means you have a **1 in 5** chance of arriving late at your destination. Of course, this depends on the airport you're flying from and to, and the airline that's taking you. According to flightstats.com, Beijing Capital Airport has the worst record for departure delays, with only 38 per cent of flights taking off on time. But Beijing Capital is a growing airport and is still adapting to the rapidly increasing amount of air traffic it has to handle.

ANYONE SEEN MY SUITCASE?

A delay in the flight can seem like light relief compared to some of the mishaps that can befall the airline passenger. What if you arrive at the baggage carousel and find that the suitcase you packed full of the latest beachwear, evening wear, sun screen, shoes and whatever else, hasn't arrived? Millions of passengers lose their luggage every year, and although 85 per cent is traced and reunited with its owner within a day or two, plenty of passengers never see their luggage again.

On average, 0.67 per cent of passengers will have their luggage lost. That means that for every flight carrying 150 passengers or more, there will be somebody standing bereft at the baggage carousel.

Is there a doctor on board?

Some people worry about being taken ill in mid-flight, and rightly so. As in-flight comforts are pared down in the name of low cost, the chances of receiving expert medical attention aboard a plane are fairly slim. But it does happen.

In 1995, 39-year-old mother of three Paula Dixon was the beneficiary of a remarkable stroke of luck on a flight home to England from Hong Kong. On the way to the airport, Paula had been knocked off the back of her friend's motorcycle, but had boarded the plane nonetheless. During the flight she suffered a collapsed lung, which threatened her life.

Her first stroke of luck was that she was sharing the flight with not one but two doctors. One was Angus Wallace, a Professor of Surgery. The other was Dr Tom Wong, himself a trauma expert. Together they decided to perform the emergency surgery that would save Paula's life.

Not having the facility of a fully equipped operating theatre to hand, they were forced to improvise. A coat hanger, sterilized in brandy, was pushed through Paula's chest into an air pocket in the lung. The air was then evacuated via a makeshift catheter and a plastic water bottle, held in place with sticky tape. The operation took just 10 minutes and there was enough brandy left for the gallant doctors to toast its success as they went back to their seats.

SOME LIKE IT HOT

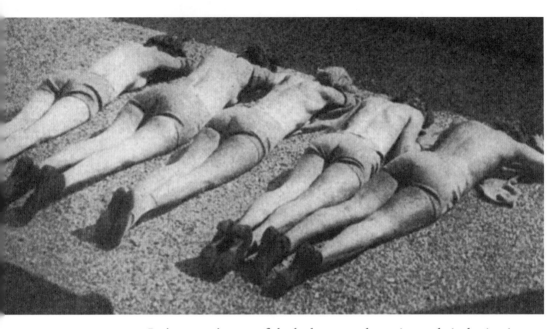

Let's say you're one of the lucky ones who arrive at their destination on time and with luggage intact; it's time to unpack and throw yourself into that dream vacation you've been looking forward to for so long.

But don't be too hasty. Sure, you want to get out there under the sun and start working on that all-over tan. But don't forget the sun screen. Despite the heightened awareness, sunburn is still a major threat. One survey found that 42 per cent of holidaymakers get sunburned every year.

Not only is it painful, it also increases your risk of melanoma, a form of skin cancer, which is now four times as common as it was in the 1970s. If you suffer sunburn five times or more, you double your risk of melanoma. Nine per cent of women and 22 per cent of men diagnosed with melanoma die within five years.

DON'T DRINK THE WATER!

It's easy to protect yourself from sunburn. A more subtle threat is food poisoning. With more than 250 different forms of food-related illnesses waiting to get you, there is always a chance of suffering an upset stomach when you go on holiday, and of course some places are more risky than others.

When you think of the countries where you might expect to come down with Delhi Belly, Montezuma's Revenge or whatever else you want to call it, the USA doesn't spring to mind. Yet **1 in 4** Americans fall ill through some food-related disease each year, **1 in 1,000** are hospitalized by it and **1 in 60,000** die. So watch what you eat!

LOVE IN A HOT CLIMATE

But why focus on all these unlikely events when you're supposed to be enjoying yourself? Holidays are a time for joy and romance, and there is statistical proof. More than **1 in 3** people have a romantic fling on holiday – that's more than the number who actually go looking for one. And while 60 per cent of those holiday romances end at the departure gate, an encouraging 15 per cent of holidaymakers begin a lasting relationship while they're away.

LOVE HURTS

Young love is a sweet and tender thing. Or it should be. But a survey of high school students in the USA found that 1 in 10 suffered physical violence from someone they were dating in a year. And it's not just girls who are the victims. In fact, the proportion of girls who reported such violence was slightly less than 10 per cent, while the proportion of boys was slightly above.

Physical violence isn't the only form of abuse dished out between dating couples. Nearly half of all teenagers claim to have been victims of some kind of controlling behaviour, be it physical or sexual abuse, threats or harassment.

Technology has brought a new dimension to the issue. Victimization by a boyfriend or girlfriend via a mobile phone, known as 'textual harassment', is reported by **1 in** 4 teenagers, and is growing.

Perhaps it's all part of the process of finding your ideal partner, but the abuse doesn't end in adulthood.

Physical violence happens to more than a quarter of all women at some time in their life – and to **1 in** 7 men.

WHO'S THE DADDY?

One of the most devastating experiences that can happen in any relationship is the discovery of a paternal discrepancy. In other words, the man believed to be the father of a child turns out not to be. It isn't always a deliberate fraud on the part of the mother – pregnancies can occur at the time of a change in partner and can lead to a genuine mistake in identifying the true father.

Historically, paternal discrepancies have tended to go undiscovered unless it has all come out in the heat of an argument. But the development of DNA testing has brought more and more cases to light, especially as it is now widely used to settle paternity disputes in divorce cases.

In England, a study by Liverpool John Moores University in 2005 came up with a figure of 3.7 per cent as an estimate of the incidence of paternal discrepancy among the general population. That's **1 in 27**

men believing themselves to be fathers when, in fact, they are not. A subsequent study at the University of Oklahoma yielded a similar figure: 3.85 per cent. Where DNA tests are carried out in order to settle a paternity dispute, the figure rises to **1 in 3**. That's a lot of mothers keeping mum!

MARRIAGE AND DIVORCE

In 1987, a notorious article in *Newsweek* stated that a single woman aged 40 was 'more likely to be killed by a terrorist' than to ever marry.

This rather tongue-in-cheek remark was based on research that showed that a single woman aged 30 had a **1 in 5** chance of marrying, and by 35 her odds had dwindled to a forlorn **1 in 20**.

That was then. Ten years later a census revealed that a 40-year-old single woman actually had a **2 in 5** chance of marrying. The fact was, people were leaving it later and later to tie the knot. And the trend continues today.

A hundred years ago, the average age of a first-time bride was 22, while for a groom it was 26. Today the average age has increased to 28 for women and 30 for men.

But it hasn't been a steady rise. Back in the 1950s and 1960s, young marriages became very fashionable and girls were getting married at an average age of 20, to husbands just two years older.

The children of that generation – the 'baby boomers' – have taken a somewhat different outlook on the race to marry. **1 in** 7 women born in those years got married after the age of 30.

Back in the Middle Ages, when it was in a father's interests to find a good husband for his daughter, most girls were married in their teens, often to husbands 10 years older or more. Today, **1 in 2** teenage marriages ends in divorce within 15 years.

As love began to play a greater part in the institution of marriage, the median ages drew closer together, and then began to rise. With more and more women going to college and building their own careers, together with a growing inclination to live together before making a commitment (**2 in 3** couples cohabit before getting married, and most live together for about three years before making a decision one way or the other), today people are getting married later than ever… and less than ever.

The marriage rate in the western world has fallen by about a third in the last 35 years. The belief in the sanctity of marriage has diminished, with **1 in 3** teenage girls saying they don't expect to be married to one man for life. They have seen the divorce rate rise to just short of 40 per cent, with the average marriage lasting just 12 years.

But while a man who remains a bachelor beyond the age of 40 is still highly unlikely to ever throw in his lot with a woman, that single 40-year-old woman can now rate her chances of marriage as almost 50-50.

Oh, and as for the fabled Seven Year Itch – it's actually more like four-and-a-half.

TRAVEL BROADENS THE MIND – AND SHORTENS THE LIFE

Approximately **1 in 2,000** lives come to a premature end each year as a result of an accident. The biggest killers are traffic accidents, which account for a third of all accidental deaths.

In western Europe, the Dutch have the most enviable safety record, with your odds of being killed on the road in any given year standing at **1 in 18,500**. That's nearly three times better than in the USA, where there are 765 cars for every 1,000 people – more than any other country in the world. In the Netherlands there are only **457 cars per 1,000** people.

However, the correlation between the number of cars per capita and the risk of dying on the road does not apply throughout the world. Some of the worst traffic accident rates occur in African countries where car ownership stands at around one per cent or less.

In the Ivory Coast, for example, car ownership stands at 1.2 per cent, yet **1 in 1,900** people die on the road each year.

But if you think that's bad, it's nothing on trouble-torn Iraq. According to the World Health Organization, **1 in 1,000** Iraqis died in traffic accidents in 2004.

They say that you stand a far greater chance of being killed crossing the road than dying in an air accident, and the statistics bear this out.

Even among the 25 airlines with the worst accident records, the odds of being killed are **1 in 843,744.**

Among the 25 safest airlines, the odds of dying in a plane crash are **1 in 9.2 million.**

However, what makes flying frightening is that if you are involved in a fatal air accident, your own chances of survival are less than **1 in 4.**

SOFT LANDING

When Captain Chesley Sullenberger successfully landed his US Airways passenger jet in the Hudson River in January 2009, preserving the lives of all 155 people on board, he defied the odds. The normal chances of surviving a plane ditching under controlled flight are just 53 per cent.

Yet, after a flock of Canada geese flew into his aircraft shortly after take-off from La Guardia airport, causing both the engines to fail, he weighed up his options and decided that the safest course of action was to try a water landing, just yards from the skyscrapers and ferries of New York City.

He managed the feat with such aplomb that even the flight attendants didn't realize they'd landed on water until they opened the doors.

AS EASY AS FALLING OFF A CLIFF

Every year, **1 in 175** people on Earth suffer a fall that requires medical attention, and more than one per cent of those falls are fatal.

This makes falling the second-biggest cause of accidental death, behind traffic accidents.

A fall can be anything from falling off a building to tripping over the cat and falling on the kitchen floor.

In 2004, 405 people in the USA died falling out of bed.

Would you let your children run free along a cliff edge? What is it about being so close to a precipice that makes you feel like you're being drawn towards the abyss? Could it be the fact that many people are?

At the last count, Japan and the USA shared the highest number of fatal falls from cliffs with 82 in a year. However, in terms of probability, that works out at a mere **1 in 1.5 million and 1.6 million** respectively.

Austria tops the table with a probability of **1 in 167,000**. That could have something to do with the proliferation of precipices in mountainous Austria. Belgium, unsurprisingly, is not listed.

So it appears you have every reason to trust yourself as you teeter

along on that picturesque coastal walk. Yet anyone or anything can stumble and fall: hikers, cyclists, dogs…

Imagine how Charles and Linda Everson of Westland, Michigan felt as they drove along a highway in Chelan County, Washington, while celebrating their first wedding anniversary, when a cow fell from the sky and crashed on to the top of their minivan. The 600lb (273kg) beast had lost its bearings and toppled off a cliff 200ft (60m) above.

What are the odds of that!

The Eversons, though inches away from meeting a beefy end, escaped shaken but unharmed. The cow, however, was less fortunate and had to be put down due to the extent of its injuries.

The fact is, you face worse odds lying in your bed than you do walking along a cliff edge. Of course, if you choose to sleep on the edge of a cliff you improve your odds of falling to your death dramatically. It is not uncommon for daredevil climbers to set up a harness on a sheer rock face and bed down for the night. In 2007, 22-year-old Adam Young from England spent seven nights on a small aluminium and nylon

hammock, bolted to the vertical face of Cheddar Gorge, 330ft (100m) off the ground.

In his defence, he did it to raise money for a cancer charity, having lost his sister to the disease.

43

GOING UNDER

Drowning competes with falls as the second most common cause of accidental death in the world. It claims the lives of nearly 400,000 people each year. But only 3 per cent of those deaths occur in First World countries. In the USA and Australia, for example, the death rate from drowning is about **1 in 100,000**. That's a 10th of the rate in Zambia, and a 20th of the rate in Ivory Coast. Ivory Coast accounts for 1 per cent of all the world's drowning deaths.

The sea is an obvious source of danger, but most drownings take place away from the sea, in swimming pools and inland lakes and rivers.

In fact, very few drowning deaths occur while swimming. Most fatalities (**1 in 5**) are the result of falling in accidentally. Alcohol accounts for **1 in 6**, swimming for only **1 in 12**.

Children under the age of 5 are the most at risk from drowning.

And over half of the global drowning fatalities occur among children under the age of 14. Four minutes without oxygen is enough to cause permanent brain damage.

Survival at sea

Drowning at sea was a major cause of fatalities during WWII. Many ships went down with no survivors. The greatest lost of life from one incident occurred when the *USS Indianapolis* was torpedoed by the Japanese submarine I-58. The ship sank within 12 minutes, taking around 300 of its crew down with it. Many of those who survived the initial catastrophe were left floating in the water, where they died of exposure or shark attacks. Only 316 of the 1,196 crew survived – just over **1 in 4**. But the Japanese losses at sea were far greater. **1 in 5** of the Japanese Navy lost their lives during WWII.

During the Battle of the Atlantic, between 1939 and 1945, 3,500 Allied merchant ships were sunk and 175 warships. In return, the Germans lost 783 U-boats. One such merchant ship was the *SS Ceramic*, a passenger liner, which left Liverpool for Sydney on 23 November 1942. She was carrying 264 crew, 14 gunners and 377 passengers, of whom 244 were military personnel. On 5 December *Ceramic* left the convoy and headed off alone. She was never seen again.

At midnight on 6 December she was holed by a German U-boat, U-515. She stayed afloat for three hours, allowing all the lifeboats to be deployed, but there was a terrible storm. The chances of survival became increasingly slim. At midday the following day, U-515 returned under orders to pick up the captain. The storm was still raging so the U-boat picked up the first person they could reach, Sapper Eric Munday of the Royal Engineers. By the time Allied vessels arrived on the scene two days later, there was no sign of any other survivors. Munday was the only one of 655. He then survived for a month on board the U-515 as she withstood depth charge attacks from Allied ships, and was landed at Lorient in France. He was transported to a PoW camp, where he saw out the rest of the war.

SOLE SURVIVORS

Being the sole survivor of a disaster is a mixed blessing. One study of plane crash survivors found that 60 per cent experienced feelings of guilt. With sole survivors, those feelings are intensified as there is no-one to share them with. There is a sense of the miraculous about any sole survival, even though there is always a statistical chance of it happening. More often than not, the survivor is left wondering, 'Why me?'

1 IN 20

That's the question that dogs Marcus Luttrell, a US Navy Seal, who was the sole survivor of a Taliban ambush in Afghanistan in 2005. Nineteen of his comrades were killed in the firefight, but Luttrell, severely wounded, managed to crawl away and hole up in a nearby village. Six days later he was rescued.

'I don't know why I survived,' he said. 'I'd put my foot one way, and step, and someone else would do the same thing and get hit. That's one of the things about being a lone survivor – you don't know why.'

Survival is almost entirely down to chance. But there are other factors that appear to influence who survives and who does not. In a plane crash, for example, it makes a difference where you sit. A study of plane crash records in the USA revealed that survival rates were significantly higher for passengers at the back of the plane than at the front. Next time a friend brags about always travelling Business Class, you might want to remind them of this fact.

Position	Survival rate
Rear	69%
Middle	56%
Front	49%

1 IN 153

Children and crew members have a statistically better chance of surviving a plane crash. They make up around 60 per cent of sole survivors.

Bahia Bakari, from Paris, was 12 years old when she survived the 2009 crash of Yemenia Airways flight 626 on its way to the Comoros Islands. All the other 152 people on board perished when the plane plunged into the Indian Ocean, making it the worst air disaster ever to have yielded a survivor. When she woke up in the water she was so confused she thought she had been sucked out of the plane by mistake, and that it had gone on and landed without her.

'I thought about my mother. I thought she must have arrived at the airport by now and be wondering where I was. She must be really angry with me, that I'd managed to fall out of the plane into the sea just before we landed.' Bahia was rescued after nine hours in the sea,

clinging to wreckage, yet it wasn't until she spoke to a psychologist in hospital that she realized her mother had been killed in the crash.

1 IN 14

As well as where you sit and how old you are, sheer determination to stay alive also plays a part. Among those who survive the initial impact, young, fit men who are regular flyers stand the best chance of escaping from the wreckage, due to their strength, stamina and knowledge of the likely escape routes.

Captain George Burk of the US Air Force refused to die after the military plane in which he was flying with 13 others decompressed in mid-flight and nose-dived into the ground. He was the sole survivor. He remembers breaking his nose upon impact and then finding himself on the ground having suffered 65 per cent burns, a fractured skull and broken spine among other injuries. Somewhere in between he found the strength to claw his way out through a crack in the fuselage.

He managed to stagger away from the burning wreckage and lie down under a tree. As he waited and dreamed of being rescued, he could feel his strength ebbing away. But he fought to stay awake and lived to tell the tale.

'The bad cop in my head told me to close my eyes and go to sleep; the good cop was saying, if you close your eyes, you're going to die. I don't want to die. Focus, focus, focus. I held pictures of my children, their mother, my parents in my head. I could have closed my eyes so easily.'

1 IN 98

Resilience and youth certainly played a part in the survival of 17-year-old Juliane Koepcke, on Christmas Eve 1971. But another major factor in plane crash survival is where you land. Soft ground, snow, haystacks all improve your odds of pulling through. But Koepcke's landing site was something of a double-edged sword.

When the Lockheed Electra in which she was flying was struck by lightning and lost its right wing, she crashed into the Amazon jungle. The jungle canopy probably broke her fall to some extent, though it didn't help any of the other 97 people on board. But now she had to make her way to civilization.

For 10 days she trekked through the jungle, evading crocodiles and swimming across piranha-infested rivers until finally she stumbled upon a lumberjack camp and rescue. Not only had she survived a free-fall from 10,000ft (3,000m), she had also survived 10 days in one of the most inhospitable environments on Earth.

1 IN 2,700

Records of air disasters between 1940 and 2008 show that a total of 118,934 people have died. Koepcke is one of 44 sole survivors during that period, defying incredible odds of **2,700 to 1**.

OCCUPATIONAL HAZARDS

Health and Safety in the workplace – the mere phrase is enough to send people to sleep on the job. For most of us, going to work involves no apparent danger, provided we can get across the street okay on our own. But spare a thought for the many workers who take their life in their hands, to some degree or other, every time they turn up on the job.

The US Bureau of Labor compiles figures showing the most dangerous occupations, and the top two spots are consistently taken by the same jobs: fishing and logging.

Fishing boats go out in all weathers in pursuit of their catch and when the sea gets the better of them, the chances of survival are slim. But it's not just the sea that poses the threat. The machinery aboard a fishing vessel is also potentially lethal if you get caught up in it.

The latest figures showed **1 in 500** fisherman killed each year in the line of work.

The logging industry actually overtook fishing as the most deadly occupation for a few years. Falling trees and branches, chainsaws, heavy machinery, the movement of huge weights of timber, the remoteness of many locations: the risk factor is high, though the latest figures showed a fall in fatalities to about **1 in 1,600**.

They say that flying is the safest way to travel, yet pilots come third in the USBL survey. Because of the relatively small number of pilots in operation, the statistics fluctuate quite considerably, but flying planes, especially small planes, is regularly up there among the most dangerous occupations. The latest figures show **1 in 1,750** pilots losing their lives each year.

Farming is a dangerous job, but not just because of the animals you deal with. Cattle need to be handled with caution but the biggest threat to any farmer is his own tractor. Tractors are responsible for 50 per cent of farm fatalities, mostly through rolling and crushing the driver. Fatigue is a major hazard in farming, increasing the threat from tractors and all the other farm machinery, as well as falls. The annual risk of death for a farm worker is **1 in 2,800**.

Falls account for more accidental deaths than any other cause, barring traffic accidents, so it's not surprising that fifth on the list of hazardous occupations is roofing. Scaling ladders all day and working in locations where one slip can lead to a severe fall, the death risk facing roofers currently stands at **1 in 2,900**.

WHEN I'M CLEANING WINDOWS

Like roofers, window cleaners take a risk every time they leave the ground. Falling off a ladder is no laughing matter, with 50 per cent of falls from three stories or more being fatal. It's the domestic window cleaner, working at lower heights, who takes the most risks, and consequently suffers the most accidents. Those who scale skyscrapers are so protected by strict precautions that accidents shouldn't happen. But they do, and when they do there's not much chance of survival.

One man who defied the odds was 37-year-old Alcides Moreno, an Ecuadorian working in New York, alongside his 30-year-old brother Edgar. On 7 December 2007, they were cleaning the windows on the 47th floor of a skyscraper in Manhattan when the platform they were standing on collapsed. The brothers fell 500ft (150m) to the ground. Edgar was killed

but Alcides somehow survived, and after intensive medical treatment, including the transfusion of 24 pints of blood and nine operations, regained consciousness and was discharged just over a month later.

Dr Philip Barie, who oversaw the treatment of Moreno, said, 'If you are a believer in miracles, this would be one.' He revealed that people who fall from more than 10 stories hardly ever survive. Barie rated the odds of anyone surviving such a fall as Moreno did as tiny, 'well under one per cent'.

POSTMAN'S KNOCK

While the chances of a window cleaner falling 500ft (150m) and surviving are remote, a postman can more or less stake his life on the likelihood that he will

get bitten by a dog at some time in his career.

Statistics from the Royal Mail in the UK show that 5,000 postmen are bitten by dogs each year – that's **1 in 10**. So any postman spending 10 years or more in the job can expect to be bitten at least once.

Despite the protestations of owners, there are millions of dogs out there that just can't be trusted. Yet you would expect a police dog to know how to behave. Unfortunately, that's not always the case, as the inappropriately named postman Joe Luckey, of Lebanon, Indiana, discovered. In August 2007, Luckey was doing his rounds when he was attacked by a 7-year-old German shepherd called Erik which worked for the police department.

The dog bit him in the face and neck, narrowly missing his jugular vein, and was suspended from duty pending retraining.

Perhaps such events are not always the dogs' fault. In Germany they have started sending postal workers on 'dog psychoanalysis' training courses, and the number of attacks has fallen by a third.

SPORT

What are the odds?

1 in 256 – an octopus correctly predicting the outcome of eight World Cup soccer matches

1 in 8,589,934,592 – losing the toss 33 times in a row

1 in 40 – simple odds of winning the Grand National

1 in 2,300 – BASE jumping to your death

80 to 1 – an Englishman winning Wimbledon by 2015

HORSES FOR COURSES

The idea of odds really comes home to us at the races. The sport of kings has been attracting gamblers for thousands of years and today the money staked on horseracing around the world amounts to around £75billion ($115billion) a year. Bookmakers have to have an intimate grasp of probability theory because they too stand to lose, and a series of losses could drive them out of business if they get their sums wrong.

So how do they calculate the odds? It can be explained most simply by looking at a two-horse race where both horses are equally matched. The mathematical probability of either horse winning is 50 per cent (evens); the probability of one or other horse winning is 100 per cent (barring any freak accidents). If the bookmaker offers even money (1-1) on both horses, and punters stake an equal amount of money on each horse, he will pay out the same amount as he has taken in: e.g. if £50 ($78) is bet on each horse, he will pay out £100 ($155) on the winner (£50 stake + £50 winnings), having taken £100 in total.

In order to make a profit, therefore, a bookmaker needs to make his total odds add up to more than the total mathematical probability. In this case, offering odds of 5-4 on (you bet £5 ($7.50) to win £4 ($6) profit) for each horse will yield a profit of £10 ($15).

Explanation

Having taken £50 on each horse, he pays out £90 on the winner (£50 stake + £40 winnings), having taken £100 in total.

The odds of 5-4 on are the equivalent of 5 in 9 (55.55 per cent), so the bookie has offered odds that total 111.11 per cent. We know that the total probability of any horse winning the race is 100 per cent.

The 11.11 per cent discrepancy is known as the overround, and this is where the bookmakers make their profit.

Calling the odds

But the job of the bookie is more complex than this. He must adjust his odds in response to the gamblers' interest. For example, if £90 is staked on one horse (the favourite) and only £10 on the other (the outsider), yet both are at odds of 5-4 on, he stands to lose heavily if the favourite wins. *Explanation*: *He must pay out £162 ($251) (£90 stake + 4/5 x £90 = £72 ($112) winnings), having taken £100 in total.*

Of course, if the outsider wins the bookmaker stands to make £82 ($127) profit (£100-£10-(4/5 x £10)). So you can perhaps see why the bookies are always smiling when the favourite loses.

Over time, however, the favourite will be expected to win more times than the outsider because it is a better horse, so the bookie will adjust the odds so that his less frequent profits from the outsider winning outweigh the more frequent losses from the favourite winning.

In real racing there will be several horses in the field but the same mathematical principles apply. And as more money is staked on one horse, the odds on that horse will be shortened (come in), while the odds on the less well-backed horses lengthen (go out).

Whenever the favourite wins, the bookies wring their hands, but when it doesn't, they rub them with glee.

THE 100-1 OUTSIDER

The Grand National at Aintree, England is the most famous horserace in the world because of its excitement and sheer unpredictability. When you back the favourite in a flat race, you're putting your money on a proven speed merchant, a horse that has been shown to like the conditions, the distance, the jockey, better than any other horse in the race, and it's all down to how it performs on the day. There are no surprise factors to influence its performance.

The Grand National is different. A huge field of 40 horses goes round two circuits of a 2.25 mile (3.63km) course, over the most fearsome jumps in racing, the highest standing at 5ft 3in (160cm). The chances of getting through all that without being bumped by other horses, unseated by the towering jumps, impeded by riderless strays or just running out of steam are slim. It's unusual for more than half the field to finish.

In the last 50 years, only one in six favourites have triumphed in the Grand National and it's very rare to see odds of less than 5-1. More often than not, the winner will go off with double-figure odds.

2008 and 2010 did see the favourites come in first, but inbetween, in 2009, a 100-1 outsider Mon Mome upset the formbook, delighting the bookies and confounding the punters. It was the first time in 42 years that a horse had defied such odds to win this famous race, but it had done so on merit, outperforming a number of more fancied horses.

For its 100-1 predecessor, Foinavon, in 1967, it had been more a case of outrageous fortune. The horse was such a hopeless bet that its owner and trainer didn't even bother to go to Aintree to watch it, and their apathy seemed justified as stand-in jockey John Buckingham cantered Foinavon round the first circuit with no apparent interest in getting near the front.

But up front, things were to take a sudden and dramatic turn. As riders approached the 23rd fence, one of the smallest on the course, a riderless horse veered across in front of them, causing chaos. Some horses refused to jump, others tried and fell, and several turned and ran back up the course in the wrong direction. Suddenly, Foinavon's lagging run looked like a canny game plan. Far enough behind to avoid the pile-up, Buckingham guided him round the melée, jumped the fence and rode to an historic, if unlikely, victory.

The 23rd fence is now named Foinavon in honour of the horse that defied odds of 100-1.

CEPHALOPOD LUCK

Paul the Psychic Octopus from Oberhausen in Germany achieved worldwide fame in 2010 for his incredible success at predicting the outcome of matches at the soccer World Cup. Paul correctly predicted the winners of all seven of Germany's games, plus the final between Spain and Holland.

Suggestions of a fix were quickly dismissed. Paul was presented with two identical boxes, each containing a mussel (his favourite food) and bearing the flag of one of the two teams in question. Anyone who tried

to influence his decision would themselves have to have had the power to predict the result. Therefore, it was indisputable that Paul was acting alone.

Paul's amazing run began when he chose to eat the mussel in the box with the German flag instead of that with the Australian flag. Germany duly beat Australia 4–0. Five days later, there were gasps of horror when he chose the Serbia box over the Germany one. Serbia were not expected to beat Germany. But Germany had a player sent off and Serbia did indeed snatch a 1–0 victory. Paul was on a roll.

He went on to predict German victories over Ghana, England and Argentina, their semi-final defeat to Spain, and their consolation victory over Uruguay to clinch third place. Then he crawled into the Spain box again to predict their ultimate triumph over Holland to become World Champions. How did he do it? There are only two rational explanations: either Paul had the power of clairvoyance, or he got lucky.

Assuming Paul wasn't psychic, his achievement was the equivalent to correctly predicting eight consecutive tosses of a coin.

$$\tfrac{1}{2} \times \tfrac{1}{2} \times \tfrac{1}{2} \times \tfrac{1}{2} \times \tfrac{1}{2} \times \tfrac{1}{2} \times \tfrac{1}{2} \times \tfrac{1}{2} = \mathbf{1/256}$$

In other words, put 256 octopuses through the same experiment and the chances are one of them will match Paul's achievement. The World Cup brings out strange behaviour in many of us, and it's likely Paul wasn't the only animal in the world challenged to predict the outcome of matches. I happen to know of a cat in Surrey who was relieved of his responsibilities after one game for failing to predict England's 1–1 draw with the USA.

So was Paul a genuine phenomenon? Or was he just a predictable consequence of probability? Unfortunately, Paul passed away peacefully in his sleep on 25 October 2010, taking his secret with him to the grave.

FIVE THINGS MORE LIKELY THAN AN ENGLISHMAN WINNING WIMBLEDON

Back in 1998, when Tim Henman was 23 and reaching his first Wimbledon semi-final, Ladbrokes were offering odds of 7-1 on an Englishman winning the men's singles title before the year 2000. We're still waiting. The odds have since gone out to 80-1 for an English win by 2015. The last time an Englishman won the Wimbledon Men's Singles title was in 1936, when Fred Perry clinched his third consecutive victory. His successor has been such a long time coming that other events have rather overtaken it.

HALEY'S COMET

The fiery space rock becomes visible to the naked eye every 75 years. When Perry won in 1936, the comet wasn't due for another 50 years and no-one imagined there wouldn't be an English winner in that time. How wrong they were. Its next appearance is due in 2061. Time is running out.

AN EARTHQUAKE

The UK isn't renowned for its earthquakes. They seldom register above 3.7 on the Richter scale and so tend to go unreported. In fact, on average Britain experiences an earthquake of magnitude 4.7 or larger every 10 years. That means there have been seven such quakes while England has been awaiting its next men's singles champion.

WAR

Three years after Perry's last win at Wimbledon, Britain declared war on Germany. And while there has been no direct military threat to mainland Britain since the Second World War, other than terrorist acts, it has continued to engage in conflicts around the world. The Iraq War brought the total to 21 separate conflicts involving British forces, including the Second World War, since an Englishman last won at Wimbledon.

A HUNG PARLIAMENT

Britain holds a General Election at least every five years and when no party achieves an overall majority, it is declared a hung parliament. There have been two hung parliaments since 1936, one following the General Election of 1974 and again in 2010. Prior to 1974, the last hung parliament in Britain was in 1929. They tend to occur in times of crisis – a word that is often applied to British tennis.

WINNING GOLD AT THE WINTER OLYMPICS

Just as tennis becomes the nation's favourite sport for two weeks in June and July, the Brits develop an overnight passion for things like curling and bobsleigh during the Winter Olympics. Since 1936 Great Britain has won a magnificent SEVEN gold medals. That averages out at one gold medal every two-and-a-half Winter Olympiads.

HITTING A HOLE IN ONE

1 IN 12,000

Every golfer dreams of hitting a hole in one but few ever get to do it. American Norman Manley claims to have hit 59 in his lifetime, while Jacqueline Gagne, a member at Mission Hills in California, claimed to hit 16 in six months!

If that doesn't make you wonder, she had only been playing the game for four years. It took her coach, Mike Mitchell, 30 years to achieve the same total.

According to *Golf Digest*, an average golfer's chances of hitting a hole in one are around **1 in 12,000**.

Insurance companies are always interested in the probability of such things occurring. One puts them at **1 in 12,750** for an average amateur and **1 in 3,756** for a pro.

And if you plan on doing it twice in the same round, the odds go out to **1 in 67,000,000**.

Website Holeinone.com keeps a database of aces occurring around the world, which has thrown up some useful stats for anyone desperate to fulfil that golfer's dream. For example, if you want to go out and hit a hole in one, the best day to choose is a Friday, the worst a Sunday.

The average handicap of a hole in one golfer is 13–14, and the average age is 44. The average yardage of an aced hole is 150 (137m) and the most successful club is a 7-iron.

So where did Jacqueline Gagne fit into all that? Well, at 46 she was around the optimum age. However, investigating journalist Dave Kindred from *Golf Digest* found a disturbing lack of evidence to support her claims, and an even more disturbing statistic to challenge them.

Kindred got mathematician David Boyum, a Ph.D from Harvard, to calculate the odds and he came up with this figure:

1 in 2,253,649,101,066,840,000,000,000,000,000,000,000.

WILL I EVER THROW A 180?

There's less luck involved in throwing a 'maximum' in darts than there is in hitting a hole in one. In golf, the conditions play a major part in the run of the ball and when you hit your tee shot, you have no way of knowing exactly what the conditions are on the green. It's hit and hope.

If you aim to throw three darts in the treble 20 and they all go in, that's not luck – it's the ultimate in sporting precision. There are no external forces in action. For those who don't play, a 180 is scored by throwing three darts together in the treble 20, a segment measuring approximately 0.62in^2 (4cm^2).

The pros do it with remarkable regularity – about **1 in 10**. For an average player, the magic maximum is achieved once in a thousand goes. This theoretical probability was published by Kari Kaitanen in *The Dart Book*. It also states that the probability for a beginner is **1 in 10,000**, although it's fair to say that after 10,000 throws, a player can no longer be deemed a beginner.

GOING FOR GOLD

On 17 August 2008, at the Olympic Games in Beijing, American swimmer Michael Phelps swam the 100m butterfly in 50.1 seconds. It was the third leg of the 400m

medley relay, and Phelps' time was a record for the event.

Not surprisingly, his team won and he claimed a record-breaking eighth gold medal for the games, beating Mark Spitz's previous record of seven.

Four years earlier, in Athens, Phelps had fallen one short of Spitz, winning six golds. His grand total of 14 made him the most prolific Olympic gold medallist ever, five clear of his nearest rival, gymnast Larissa Latynina of the Soviet Union.

1 IN 36

1 in 36 were the simple odds of winning one gold at the Beijing Olympics, if you calculate them based on the number of athletes and the number of medals awarded. Of course, the figure varies according to the number of competitors in your event and the number of events you enter.

Competing for the medals is, you could say, the easy bit. The odds of becoming an Olympic athlete in the first place are far longer. The

11,000 athletes who took part in Beijing represented 1/600,000th of the world population. So the odds of becoming an Olympic athlete and winning gold at the 2008 Olympics were roughly **1 in 22,000,000**.

But what are the odds of becoming a gold medallist? The minimum age to compete in the Olympics is 14 (although the youngest-ever gold medallist was Marjorie Gestring of the USA, who was 13 years and 268 days when she won the Springboard Diving event in 1936). The oldest gold medallist was Oscar Swahn of Sweden, who was a member of the winning Running Deer, Single Shot team in 1912 at the age of 64. So if we take the ages of 14 to 64 as the scope of an athlete's competitive life, that's a span of 50 years, or potentially 13 Olympiads.

According to the UN, world population is forecast to grow to 9 billion over the next 50 years, giving an average of 7.8 billion. Anyone aged 14 at the next Olympic Games will have a possible 13 Olympics to compete for, giving a total of around 3,900 gold medals.

Therefore, the odds of winning gold in a lifetime for someone yet to compete at their first Olympics are currently:

3,900 in **7,800,000,000**   or 1 in **2,000,000**.

WEIGHING UP THE DANGER

The great racing driver Juan Manuel Fangio once said, 'A crazy man finishes in the cemetery.' But his great rival Stirling Moss offered his own assessment. 'To achieve anything in this game, you must be prepared to dabble on the boundary of disaster.'

Today the odds of losing your life in a Formula One race are practically zero, but that's a far cry from the days of Fangio and Moss. Motor racing is a sport of thrills and spills, with much of its appeal derived from its sense of danger. The deaths of racing legend Ayrton Senna and Roland Ratzenberger in the same race weekend at San Marino in 1994 heralded a period of unprecedented safety on the track and anyone coming to Formula One racing for the first time today, seeing the way drivers crash at speeds in excess of 150mph (240 km/h) and walk away unscathed, might find it hard to believe that this sport once claimed lives at a frightening

rate. Yet Senna was the 45th driver to die on the track, in racing, practice and testing, since the F1 World Championship began in 1950. His death took the rate to one fatality per year.

Senna was the 24th driver to die while racing, but it came as a shock – there hadn't been a driver killed in a race since Riccardo Paletti at the Canadian Grand Prix 12 years before. Safety standards had improved no end since the 1970s, when seven drivers had lost their lives during competition.

Risking your life was part of the appeal of motor racing, a sport that threw up heroes prepared to go wheel-to-wheel regardless of the odds. And every year they were reminded of the risks they were taking.

WINDY AT THE INDY

The 1950s saw drivers dying at a rate of three every two years. The circuit to avoid was the notorious Indianapolis Motor Speedway, home to the

Indianapolis 500, which was part of the Formula One World Championship in the 1950s. Between 1953 and 1959, Indianapolis alone claimed drivers' lives at a rate of one per year. In Europe, the track that struck fear into the hearts of racing drivers was the Nürburgring in Germany. A narrow, undulating circuit that wound through a forest, the 'Ring' was no place to lose control. Five F1 drivers perished there between 1950 and 1970, and even though the fatalities stopped in the 1970s, the accident that nearly claimed the life of Nikki Lauda was a stark reminder of the mortality of these gods of the racetrack.

DANGEROUS SPORTS

The dangers involved in Formula One motor racing are nothing compared to motorbike racing. The deadliest bike event of all is the Isle of Man TT meeting, run over the perilous Snaefell Mountain course in the UK. Between 1907 and 2010 the course claimed 229 lives – a rate of 2.2 per year.

Some sports appear to have no other purpose than to defy death. BASE jumping – the practice of parachuting from a fixed object (a cliff or a building, say) – is the ultimate leap of faith: faith in one's parachute. With BASE jumping, it's not so much the height they jump from as the lack of it. Jump off a 500ft (150m) building and you've got less than six seconds before you hit the ground. You'll be falling slower than in a normal jump from a plane, so your 'chute will have less force pulling it open. And if you're lucky and it does open, you only have a few seconds to choose your landing site.

In 1990 BASE jumper Russell Powell jumped from the Whispering Gallery inside St Paul's Cathedral, London, and lived to tell the tale. At 112ft (34m), it stands as the lowest BASE jump ever completed.

Powell's philosophy, that 'a long life might not be good enough, but a good life is long enough,' might yet prove prophetic. According to research, BASE jumping is by far the riskiest sport, with a death rate of **1 in 2,300** jumps. Skydiving, by comparison, has a fatality risk of about **1 in 100,000** jumps.

A lot of so-called 'dangerous' sports actually carry a lower risk of dying than more popular sports, such as swimming, cycling and running, all of which have a higher death rate than skydiving.

Cyclists and runners get hit by cars, swimmers drown. The physical condition of many of those who undertake these popular sports may not always be up to the strain.

Lawn bowling has one of the highest fatality rates of any sport, but one has to ask, would the majority of players, getting on in age as they are, stand a better chance of survival if they stayed at home and watched TV? Unlikely.

Those who go in for dangerous sports tend to be young and fit, protected by careful safety measures and not vulnerable to external threats, such as cars. Hence, the odds against biting the dust are better than you might imagine.

Fatalities in dangerous sports	
Hang-gliding	1 in 116,000 flights
Scuba Diving	1 in 200,000 dives
Rock Climbing	1 in 320,000 climbs
Skiing	1 in 1,556,757 visits

HEADS I WIN, TAILS YOU LOSE

Tossing a coin is the ultimate expression of unbiased decision-making, the accepted wisdom being that a coin spins randomly and, therefore, the odds of it coming up heads or tails are both **1 in 2**.

However, experiments into coin tossing, carried out in laboratory conditions, with the coin being tossed by a mechanical device that imparts exactly the same upward and rotational force every time, have proven that a coin being tossed in this way will spin exactly the same number of times every time.

The randomness comes from the human element: the inconsistent force exerted in tossing the coin. If you can learn to make your coin tossing consistent, you can always control the outcome of the toss.

But is it worth it? The coin toss is most commonly used at the start of sports matches to decide who begins the game. But what bearing does winning the toss have on the outcome of any game? Of the 43 Superbowls that took place up to 2010, the team that won the toss was victorious only 20 times. In other words, it was better to lose the toss.

In cricket the toss is deemed to be far more significant, because playing conditions change during the course of the game. However, a study by Basil M de Silva and Tim B Swartz, which looked at the statistics from 427 one-day internationals played in the 1990s, concluded that, 'contrary to widespread opinion, winning the coin toss at the outset of a match provides no competitive advantage'.

UNLUCKY CALLS

In 2010, 16-year-old cricketer Kirsty Perrin of Macarthur, Australia, completed three seasons of luckless coin tossing. Despite varying her strategy, she racked up 33 consecutive losses. The odds of that happening are **1 in 8,589,934,592**, roughly a thousand times more unlikely than winning the Oz Lotto Australian lottery.

TOAST FLIPPING

It's not exactly a sport, but the acrobatic performance of toast falling through the air is something that has intrigued mankind for years. Why does it always seem to land butter side down?

An irritating truth

There have been countless experiments carried out to test this phenomenon and the overwhelming evidence is that a piece of buttered toast dropped from table height will land butter side down most of the time (95 times out of a hundred, according to my own experiment).

There is a simple explanation for this. A piece of toast, when dropped, begins to rotate as it falls. The rate of rotation is determined by the size of the toast and the uneven distribution of weight (the butter makes one side heavier). Therefore, as toast tends to be a standard size, we can say that all buttered toast rotates at the same speed. The height of a table top is also fairly standard and this happens to be high enough for the toast to make just a half rotation before landing. Therefore, it is almost inevitable that a piece of toast dropped from table height will land butter side down.

But sometimes our efforts to catch the toast as it falls impart an extra rotational force that brings the toast round 360 degrees before landing. There are ways to counteract this rule. If you make the buttered side of the toast concave by pressing firmly as you spread it, this will affect the way the toast falls, bringing it to land butter side up more than 50 per cent of the time. Another safeguard is to begin with the toast butter side down, or, better still, not to drop it in the first place.

MONEY

What are the odds?

1 in 660 – becoming a millionaire

1 in 5 – living in extreme poverty

1 in 14 million – winning the lottery

1 in 66,666 – getting a straight flush in poker

1 in 1 million – finding a natural, gem-quality pearl

WHO WANTS TO BE A MILLIONAIRE?

So you want to be rich? You want to have a million to call your own? What are the chances?

First, let's establish our currency. A million rupees is far less impressive than, say, a million pounds. In order to keep things impressive, we'll use the US dollar.

There are about 10 million US dollar millionaires in the world. That's **1 for every 660** people.

So what's your best chance of becoming one? Forget the national lottery and the slot machines, if you want to know what it feels like to really feel a million dollars, set up your own business and work hard. That's how 99 per cent of millionaires make their money.

And the best place to do it? America, of course. North America is home to almost a third of the world's millionaires, with **1 in 160** people feeling the comfort of the cash.

Asia has now overtaken Europe as the second most fertile breeding ground for millionaires in terms of overall numbers, and is on its way to overtaking North America. But because of the huge population, your odds in Asia are ten times worse than they are across the Pacific.

TO HAVE AND HAVE NOT

The sad truth about wealth is that the vast majority of it is in the hands of a tiny number of people, leaving the vast majority of the world's population with very little at all.

A study at the turn of the millennium found that 40 per cent of the world's wealth was owned by just one per cent of the adult population. The flip side of this sorry statistic was that half of the world's adult population were sharing a mere one per cent of the world's wealth.

The proportion of the world's population living in extreme poverty (on less than 81p ($1.25) a day) has halved in the last 30 years, but the total number is still alarmingly high: 1.4 billion. That's more than **1 in 5**, distributed throughout the world.

If you're reading this book, the chances are that you're not one of those **1 in 5**. It is a matter of chance that some of us are born into poverty

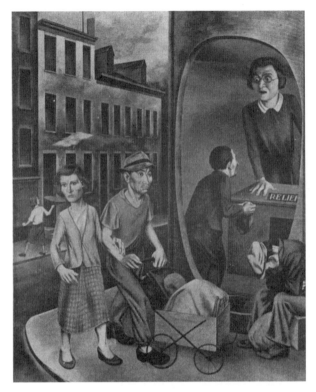

and some of us are not. In South Asia, the odds of living in extreme poverty are **2 in 5**. In sub-Saharan Africa, they are **1 in 2**.

Those born in the western world are more lucky, though the discrepancy between the haves and have nots is still stark. In the USA, **2 in 5** people share just one per cent of the nation's wealth, while **1 in 10** enjoy the trappings of 70 per cent of it.

Having a roof over your head, a place to call your own, is one of the fundamentals of civilized society – but with the price of property and the demand for affordable housing, there is always the fear in the back of the mind that we could lose it all and end up on the street.

The sub-prime crisis and recession in the USA brought this very issue home to roost. Between 2007 and 2009, the number of families seeking shelter in America rose by 30 per cent.

It brought the total of homeless individuals in the USA to 1,035,000 – **1 in every 300** head of population. And of that number, 383,000 (just over one third) were part of a homeless family.

So you find yourself and your children out on the street: what are your chances of being given emergency shelter for the night? Of those homeless families in America, **1 in 5** were forced to sleep under the stars on any given night.

WINNING THE GREEN CARD LOTTERY

Each year, millions of people apply to become citizens of the USA. And each year the USA grants 50,000 Green Cards under the Diversity Visa Lottery Program, commonly known as the Green Card Lottery.

In 2009 there were 9.1 million eligible applicants. That number was rapidly whittled down to just 99,600 who were invited to apply for one of the visas.

90 out of every 91 eligible applicants were rejected.

The remainder were entered in the lottery, which is drawn completely at random. Their odds of winning were just under **1 in 2**.

But of all eligible entries, the odds of winning the Green Card Lottery were **1 in 182**.

America is often accused of being an insular country, and the argument is backed up by pointing out that most people in the country don't own a passport.

This is true. To an extent. According to recent figures, only 29 per cent of Americans (**1 in 3.45**) are in possession of a passport. However, this number has risen by more than five per cent in the last decade.

Contrary to popular opinion, there's a growing desire among Americans to get out and see the world.

IT'S A LOTTERY

So you've decided to disregard my advice and put your faith in the lottery to make you a millionaire. Okay, let's take a look at your odds.

Don't be fooled into thinking your chances of winning the lottery depend on how many people take part. That only affects the amount of money you win, since the prize pot will have to be divided up if more than one person wins.

The odds of drawing the winning numbers remain the same regardless of how many people play, because they are dictated by the number of possible outcomes in the draw.

The usual format for a lottery is to have X possible numbers from which Y are drawn. In order to win the jackpot you have to match all

six numbers drawn. The odds of doing so are calculated by this rather frightening looking formula:

X!/(Y! (X-Y)!) where X is the total number of balls and Y is the number of balls drawn.

The ! stands for factorial, which means the product of that number times all the numbers smaller down to 1, multiplied together.

For example, **5! = 5x4x3x2x1 = 120**

So, in the case of a lottery in which six balls are drawn from 49, the formula reads:

49!/(6! (49-6)!)

Take it from me, the answer is 13,983,816. So you have a **1 in 13,983,816** chance of winning the jackpot.

The game is given more interest by offering further prizes for matching fewer balls, and throwing in a bonus ball. Second prize requires you to match five of the first six balls, plus the bonus ball. You may wonder why the odds for this are not the same as for the jackpot. After all, you're still matching six numbers from 49.

The difference is that you have six different ways of doing so, because of the addition of the bonus ball. Therefore, the odds are **13,983,816/6 = 1 in 2,330,636**.

The remaining prize odds are as follows:-

5 numbers   **1 in 55,491.33**

4 numbers   **1 in 1,032.4**

3 numbers   **1 in 56.7**

Note, this doesn't mean that **1 in 56.7** people will match three numbers. It means that any individual can expect to match three numbers once every 56.7 goes. Is it worth it?

WHY THE HOUSE ALWAYS WINS

When you gamble your money at the casino, it's helpful to know the odds. It's also wise to understand that the odds are biased in favour of the house.

With roulette, it's all in the zeros. The two slots for 0 and 00 were included to give the house a mathematical advantage. They brought the total number of slots into which the ball could fall to 38, yet these extra slots were discounted when calculating the payout. Only the numbers 1 to 36 were taken into account, so, for example, a bet on Red was deemed to have an **18 in 36** chance (**1 in 2**) and, therefore, earned a payout equal to the stake (evens).

In 1843 in Homburg, Germany, a new roulette wheel was introduced with just one extra slot for 0, reducing the total number to 37. Omitting the 00 was a business move to gain a competitive edge over rival casinos by making the wheel more attractive to gamblers. This wheel became the standard in Europe and remains so today, whereas American roulette wheels still have a 0 and a 00. Therefore, you have a better chance of winning in Europe than you do in the USA, as the table opposite shows.

But the house still has the advantage.

ROULETTE ODDS VERSUS THE HOUSE LIABILITY

Bet	European	American	Payout
Straight Up (1 No)	36 to 1	37 to 1	35 to 1
Split (2 Nos)	17.5 to 1	18 to 1	17 to 1
Street (3 Nos)	11.334 to 1	11.667 to 1	11 to 1
Corner (4 Nos)	8.25 to 1	8.5 to 1	8 to 1
Five Nos	6.4 to 1	6.6 to 1	6 to 1
Column (12 Nos)	2.083 to 1	2.167 to 1	2 to 1
Dozen (12 Nos)	2.083 to 1	2.167 to 1	2 to 1
Red or Black (18 Nos)	1.056 to 1	1.111 to 1	1 to 1
19-36/1-18/Even/Odd	As above		

The average return on a slot machine is 93 per cent.

In other words, for every pound you put in, you win 97p. Another way to achieve this is to take 3p and drop it down a drain.

PLAYING THE ODDS

Blackjack is one of the most popular casino games because, with a little knowledge, it is possible to restrict your losses to less than one per cent. A system known as 'basic strategy' will teach you how to play your cards in accordance with the dealer's 'up' card (each player is dealt one card face down and one face up).

Basic strategy has been devised by calculating the odds of turning up any number in any given situation. By following basic strategy, you are merely playing the odds.

Armed with your basic strategy, you're ready to take the house on. First, check the rules at the table. There are many rule variations but the most common one applies to the dealer on 17. In some casinos the dealer is obliged to 'hit' (take another card) on 17, in others he is obliged to 'stand'. Where he has to stand, the house edge is decreased by about 0.2 per cent.

The next thing to ascertain is the number of decks being used. The house advantage increases with more cards in play.

The house advantage comes from the fact that the dealer plays last, but there are two advantages a player has over the dealer. Firstly, a player may stand on any number between 12 and 16 (a stiff hand), whereas the dealer is obliged to hit. Secondly, a player wins 3 to 2 for getting a blackjack, whereas he only loses at evens when the dealer gets one.

Therefore, the greater the probability of getting a stiff hand or a

Stiff hand probability with number of decks		Blackjack probability with number of decks	
Decks	**Probability**	**Decks**	**Probability**
1	38.76%	1	4.83%
2	38.61%	2	4.78%
4	38.54%	4	4.76%
5	38.52%	5	4.75%
6	38.51%	6	4.75%
8	38.50%	8	4.75%

This shows how the odds of getting 12–16 decrease as the number of decks increases.

This shows how the odds of getting a Blackjack decrease as the number of decks increases.

blackjack, the more the game is biased towards the player (or rather, the less it is biased towards the house).

A good way to win money is to recognize when you're in a strong position and double your stake. For example, if your first two cards add up to 11, you should double because, with a total of 16 cards in the pack having a value of 10, you're more likely to get 21 than any other total.

The odds of getting a 10 when you hit on 10 or 11 also decrease as the number of decks increases.

KNOW WHEN TO HOLD 'EM, KNOW WHEN TO FOLD 'EM

Thanks to the internet, the number of people who play poker has exploded, with the online form of the game making close to £1.9billion ($3billion) a year. Top players make millions of dollars, but far more people are happy to lose just for the excitement of playing the game.

Make no mistake, poker is a game of chance, but it's also a game of skill. The skill lies in knowing your chances. As with any card game, there is a fixed probability for every outcome, but the complexity of poker makes calculating that probability an art that is far from easy to grasp.

First, let's explain how the game works. The aim of the game is to end up with the strongest hand. There are eight different winning hands, ranging from a straight flush (five cards of the same suit in consecutive numbers) down to a pair (two cards of the same value).

If no-one holds one of these hands (having nothing is known as no pair), the player with the highest card wins.

Each player is dealt two cards face down. Based on the strength of these cards, the players bet. Once everyone has bet, three more cards are laid down on the table. This is called the flop. Any of these cards can be used in conjunction with the cards in each player's hand to make up their overall hand. They bet again. Another

card is laid. This is called the turn. They bet again. A final card is laid. This is called the river. After the river is laid, the players reveal their cards and the one with the strongest hand scoops the pot.

The value of the winning hands is determined by their odds, as follows:

Straight flush	0.0015%	Three of a kind	2.1%
Four of a kind	0.024%	Two pair	4.75%
Full house	0.14%	One pair	42%
Flush	0.2%	No pair	50.12%
Straight	0.39%		

However, the odds change as you play. For example, if you reach the river with two diamonds in your hand and there are two diamonds and two spades on the table, you only need another diamond for a flush. There are 46 cards you haven't seen, of which nine will give you a winning hand. Therefore, your odds are shortened to **9 in 46**, or just over **4 to 1**.

Now, if a heart doesn't come up in the river, you have nothing. So you have to decide whether it's worth gambling on your **4 to 1** chance.

The simple way to decide is to compare these odds with your pot odds. Your pot odds are simply the amount you have to bet to 'call' (or match the highest bet) compared to the value of the pot.

In the case described, your hand odds are **4 to 1**. Say there's $50 (£32) in the pot and you're being asked for $10 (£6.5) to call. The pot odds are **5 to 1** and, therefore, it's worthwhile taking the risk. Why? Because the pot odds show that you need to win one in six times in order to break even (5 you lose, 1 you win), while your hand odds of **4 to 1** show that you can expect to win **1 in 5** times. The odds are in your favour.

It's a long-term principle that doesn't just apply to poker – it applies to investments too.

The explanation so far assumes that you have no idea what cards your opponents are holding. This should, of course, be the case, unless your opponents are careless or you're a cheat. But good poker players learn

to read each other's habits and mannerisms.

They will learn what sort of hands their opponents like to bet on and when they like to play it safe. They will recognize how their expression changes when they have a strong hand. Every little piece of information like this affects the odds and the great players will take all this into account when making their calculations.

LET'S MAKE A DEAL

Some people try to get rich by winning game shows. There's certainly a lot of money up for grabs these days, but you have to be aware that the odds are stacked against you, just as they are in a casino. Even when they seem straightforward, there's usually a catch. A famous example of this is what has become known as the Monty Hall Paradox.

In 1963, NBC started broadcasting a game show called *Let's Make a Deal*. It was presented by Canadian Monty Hall. The show involved Hall offering deals to the contestants, who had to decide whether the deal he was offering would be beneficial or not. One of the problems involved three doors, behind one of which was the star prize. This gave rise to a probability problem, which has baffled and even outraged mathematicians ever since.

The Monty Hall Paradox works like this. The contestant chooses a door. The host then opens one of the other two doors, which he knows to conceal a booby prize. The contestant is then invited to choose either to stick with his or her first choice or switch to the last door.

Knowing that one door conceals the star prize and the other the second booby prize, it appears to be a simple 50-50 decision. But it isn't. The fact is your odds of winning improve to **2 in 3** if you choose to switch.

Here's why:

*The probability of your first choice being correct is **1 in 3**. Therefore, the probability of it being one of the other two doors is **2 in 3**. Once the host has opened one of those doors, you know that its probability of concealing the star prize is 0. Therefore, the other door must have a probability of **2 in 3**.*

So you're faced with a simple choice: stick with your door with its **1 in 3** chance of success; or switch to the other door and double your chances?

Switching wins… well, two times out of three.

FOXED BY SOCKS

Another probability puzzle that leaves a lot of people foxed concerns the problem of matching socks.

You're getting dressed in your bedroom when suddenly a fuse blows and the lights go out. All you've got to do is put your socks on, but all your socks are loose in a drawer and you can't see to find a matching pair. You know you've got 4 green socks and 4 brown socks in there. If you want to ensure that you have a matching pair when you get out into the light, what is the minimum number of socks you need to take from the drawer? (See bottom of page for answer.)

If you just take a chance and grab two socks, the odds of them being a

matching pair are **3 in 7**. Once you've taken the first sock, there are only three of that colour left out of the remaining seven socks.

If you keep more socks in your drawer, your odds improve. Say you have 10 of each colour: then the chances of drawing a pair increase to **9 in 19**, compared to **9 in 21** with four of each.

Answer: 3. The first two may not match, but the third must match one of them.

THERE BE TREASURE

We all dream of finding treasure, but is it worth going out and looking for it?

The 49ers of the California Gold Rush thought so. In 1849, as news spread of James Marshall's chance discovery the year before at Sutter's Mill in Sacramento Valley, they flooded in from all over the world. Around 85,000 prospectors arrived in California in that year alone, a number that swelled to around half a million over the entire course of the gold rush. Yet a tiny percentage of those hopeful souls struck it rich.

1 in 5 of the 49ers died within six months of their arrival, due to the harsh conditions, the spread of disease and the violent quarrels that broke out over finds and claims. Of those who survived, the vast majority returned home as poor, if not poorer, than when they arrived.

They would have been sickened to hear about Kevin Hillier, of Victoria, Australia, who, 131 years later, went out for a stroll with his new metal detector and unearthed a gold nugget weighing 60lbs (27.21kg). The colossal nugget, shaped a like a hand, was buried a few centimetres beneath the surface, almost as if it was just waiting there for Hillier to find. It was named The Hand of Faith and sold to the Golden Nugget casino in Las Vegas, where it remains, the biggest intact gold nugget in the world.

The lucky find netted Hillier over $1million (£650,000).

DIAMONDS ARE FOREVER

Scientists are always trying to improve the odds of finding diamonds, but the most famous diamond mines have been discovered by chance.

The vast Kimberley Mine in South Africa began with a 15-year-old shepherd boy named Erasmus Jacob picking up a sparkling 'pebble' near the Orange River in 1866 and giving it to his father, who sold it. The 'pebble' turned out to be a 21.25 carat diamond.

PEARLS

Pearls are formed naturally by about **1 in 10,000** oysters. Of all the pearls used in modern jewellery, only five per cent are unfarmed. The chances of finding one of gem quality are said to be **1 in a million**. It's impossible to put an exact figure on it. Suffice it to say it's extremely rare.

So imagine the odds of a man name Moule (French for mussel) finding a pearl in his lunch. That's what happened to Geoff Moule, 77, from Dorset, England, as he tucked into his starter at the Crab House Café, his local seafood restaurant in Wyke Regis.

'I have been going there every Sunday for two-and-a-half years and I always have half a dozen oysters as a starter with my lunch,' said Moule. 'To find a pearl is really quite something,' he added with classic English understatement.

The pearl was valued at £90 ($140) – enough for 10 of his favourite starters.

ARE YOU COVERED?

One industry that relies heavily on getting the odds right is insurance. It is the job of the insurance actuaries to calculate the degree of risk in any policy. They do this by looking at the statistics for similar 'exposure units' (the person or thing insured) and categorizing each case in terms of risk.

Based on these calculations, they work out the cost of your insurance premium. If, for example, you want to insure your car for £10,000 and the actuaries calculate that among all the people of your age, sex, employment, residential status etc, **1 in 10** writes off a car each year, they will need to charge a premium of £1,000 per year just to cover their liability.

In other words, the statistics tell them that if they insure 10 people like you, they are likely to have to pay out £10,000 once each year. Therefore, they need to take in £1,000 from each of the 10 insured in order to cover the payout.

Of course, insurance companies aren't in the business of merely covering their losses. They will build in their overheads and profit on top of that basic amount, so your premium will be a fair bit higher than the simple ratio of calculated risk to payout.

As such, insurance companies are effectively offering you odds, just like a bookmaker, except in this case you are betting on something that you hope won't happen.

Like bookmakers, insurance companies don't limit themselves to just the run-of-the-mill risks, like car, home and life insurance. If you're prepared to pay the premium, they will cover anything.

In 2004, Lloyds of London, which has been selling insurance since 1688 and leads the way in unusual insurance policies, provided the cover for SpaceShipOne, the first private manned aircraft to go into space, funded by Microsoft co-founder Paul Allen. The total cover amounted to $100million (£63m).

Before the 2010 World Cup, soccer fan Paul Hucker took out insurance against England getting knocked out in the first round. The cover, worth £1million ($1.55m), was actually for the mental trauma that Hucker expected to suffer in such an eventuality. British Insurance offered it to him on the conditions that five sports commentators testified that they thought England's exit to be premature, and that Hucker could provide medical evidence of his mental suffering. The premium was £105 ($162), which means that British Insurance rated the risk at less than **1 in 10,000**. They were right. Just. England went out in the 2nd round.

Personal accident cover is a major strand of the insurance business. Everyone needs their body, but some of us rely on certain parts more than others. A bricklayer, for example, is more dependent on his hands than, say, his ears. Therefore, when you take out personal accident insurance you can specify certain parts of your body that you want to take a higher proportion of the total cover.

In the 1940s, Lloyds led the way when they insured Hollywood film star Betty Grable's legs for $1 million (£650,000). Other stars followed suit, including, most famously, Marlene Dietrich. Leg insurance is common in the entertainment business. In recent years, as an indication of how celebrity earnings have soared since Betty Grable's day, singer Mariah Carey insured her legs for $1 billion (£650m). But

it's not just the ladies who need it. Michael Flatley, of *Riverdance* fame, insured his pins for $40 million (£25m), and the great Fred Astaire reportedly insured his legs for $116,000 (£75,000) each – a strange way to break it down, since the loss of one leg would, you'd think, render the other one redundant for dancing.

Moving upwards, Dolly Parton's famous breasts are insured for $600,000 (£387,000). Whether that's $300,000 each is unsure.

Back in the 1960s, The Beatles

had their fingers insured for a paltry £200,000 ($310,000). Had Lennon or McCartney not been able to pluck the tunes for *Sgt Pepper's Lonely Hearts Club Band*, the loss of earnings would have been considerably more than that. Rolling Stone Keith Richards redressed the balance somewhat with $1.6million (£1m) worth of cover for his famous fingers.

Other unusual (but understandable when you think about it) body part insurance policies include voices: both Bruce Springsteen and Rod Stewart have allegedly insured their gravelly tones for something in the region of $6million (£3.8m). America Ferrera, the actress who became famous playing Ugly Betty, insured her teeth for $10million (£6.5m).

Every entertainer has their prize asset, but topping them all is David Beckham – soccer player, celebrity husband, model, ambassador – whose $100million (£65m) insurance policy topped anything taken out by any other sportsman in the world at the time.

Cashing in

But it's not just celebrities who cover themselves. Professional coffee taster Gennaro Pelliccia insured his tongue for $13million (£8.3m). And Bordeaux winemaker Ilja Gort insured his nose to a value of £3.9million ($6m). One thing insurance companies have to assess when offering these policies is the likelihood of the holder cashing in by inflicting deliberate self-mutilation. In a grizzly parallel to burning down a building for the insurance money, there have been cases of people blinding themselves and cutting off their hands so that they could claim. And when it's only your hair that's insured, you would think it a very tempting proposition. Nevertheless, American football star Troy Polamalu managed to get his fulsome locks covered to the tune of $1million (£650,000).

ACHIEVEMENT

What are the odds?

1 in 2 – making a successful attempt on Mount Everest

1 in 1,000 – throwing a 180 in darts

1 in 5,600 – having a book published in any given year

1 in 8,031,810,176 – a chimp typing 'Macbeth'

WHAT ARE YOUR CHANCES OF PASSING AN EXAM BY GUESSWORK?

For most of us, the best chance of achieving anything in this world is to get ourselves a qualification. Hey, that's an achievement in itself! Not all of us are cut out for the pressures of exams, though. Either we forget it all as soon as we enter the exam room, or we don't put the work in to learn it in the first place.

But there is one type of exam that always offers us hope that, if chance is on our side, we might just get away with it. I'm talking about the multiple-choice exam. We've all wondered what would happen if we went into a multiple-choice exam and just picked out answers at random. You probably know someone who's done it. And there will be cases where people have passed an exam this way. But if you're thinking of giving it a try, take a moment to examine the odds.

Say there are 100 questions, each with a choice of four answers, only one of which is correct. And let's say that the pass mark for this exam is 50 per cent. That means you need to get every other question correct: **1 in 2**.

For each question, the likelihood of getting a correct answer is **1 in 4**. You might look at these figures and think you only need a bit of luck to improve your odds from **1 in 4** to the **1 in 2** you need to pass. If so, let's hope you're not taking a maths exam. Because, in actual fact, the likelihood of getting 50 or more correct answers out of 100 is… **1 in 15 million!**

99 PER CENT PERSPIRATION

Thomas Edison evaluated the make-up of a genius as 1 per cent inspiration and 99 per cent perspiration. But other people try to measure genius by way of intelligence tests.

So how intelligent are you?

The odds that you have an IQ of 100 or higher are **1 in 2**. That's because 100 is the intelligence quotient that marks the mean for all mankind.

By definition, half the population is above average and half is below.

But an IQ of 100 is not enough to earn the title 'genius'. Far from it, in fact. You need an IQ of 130 or more to qualify for Mensa, the world's biggest society for boffins. Your odds are **1 in 50**. But even that doesn't make you a genius.

So what does? Well, that's hard to say because the average genius (if that makes sense) is smart enough not to take an IQ test. Einstein's IQ has been estimated at 160; the odds of matching that are **1 in 30,000**.

If you achieve an IQ of 172 or more, then you're **1 in a million**. That's the criteria for becoming a member of the Mega Society, founded in 1982 by American philosopher Dr Ronald K Hoeflin. By those standards, 6,700 people in the world can qualify for membership of the Mega Society. Oddly enough, only about 30 have bothered.

GET AN EDUCATION

Being a genius may not be high on everyone's agenda, but the world is getting educated. In the last half century the global literacy rate has risen from 60 per cent to 84 per cent. In the developed world 99 per cent of the population can read and write. Compare that with 300 years ago, when two-thirds of the population of Europe was illiterate. Even in the mid-19th century, **1 in 3** people were unable to sign their name.

Nevertheless, living in the First World doesn't guarantee you a good

education. **1 in 10** Americans are classified as having poor literary skills, while in Canada the figure is close to **4 in 10**.

By contrast, the USA offers the best chance of higher education. 73 per cent of Americans go on to college or university, compared to just 0.3 per cent in Malawi, at the other end of the scale. In most African countries, the odds of getting a tertiary education are worse than **1 in 20**; in China they are **1 in 13**, in India **1 in 10**.

Finland, Norway and Sweden rival the USA for the availability of higher education, but in most of Europe the odds of going to college after leaving school are **1 in 2**.

MADE IT, MA! TOP OF THE WORLD!

A 2006 study of Everest expeditions showed that 10,094 people had tried to climb the world's highest mountain since 1922, of whom 2,972 had been successful – a rate of just under **3 in 10**.

More than **1 in 50** died in the attempt.

But your chances of success are improving all the time. Back in 1924, when Mallory and Irvine made their ill-fated attempt on the summit of Mt Everest, the numbers of people attempting the climb were tiny. With Tibet only opening its borders in 1921, the ascent of the world's highest mountain was a new concept.

It took until 1953 for Edmund Hillary and Tensing Norgay to make the first successful climb to the peak, by which time 13 people had already lost their lives on the mountain.

Climbing Everest was, it's fair to say, no easy task. Today, however, hundreds of people each year arrive at the two base camps, one on each side of the mountain, to try their luck. According to the records kept by the Himalayan Database – The Expedition Archives of Elizabeth Hawley, 2009 saw 435 climbers ascend above Base Camp, of whom 229 made it to the top.

That ratio was slightly down on the previous two years, when the proportion of ascents to attempts was **1 in 1.7**. But it maintained a trend for the past decade that has seen the odds of making it to the top settle at an average of **1 in 2**.

FIRST AMONG EQUALS

'Some are born great, some achieve greatness and some have greatness thrust upon them.'

So wrote William Shakespeare in *Twelfth Night*, as he took another major step on his own way to achieving greatness.

There's little point considering the odds against being born great or having greatness thrust upon you, as there is not much you can do to effect either. Not everyone aspires to greatness, and many consider themselves unlikely to ever get the break. But recent history has thrown up plenty of leaders who have shown that greatness, in the leadership sense, can be achieved against overwhelming odds.

Barack Obama was not born great – not by any means. Being the son of a Kenyan, the odds were stacked against him ever becoming President. Not only was he the first African American to achieve that office, he was also the first President born in Hawaii. But they do say anyone can become US President. Actually, there are certain stipulations: you have to be a natural-born US citizen; you have to be over 35 years old; and you have to have lived in the US for 14 years. The natural-born US population currently stands at about 270,000,000, of whom roughly half are under 35. The average life expectancy is 78, which means the average citizen has 43 years in which they are eligible to become US President. If there is an election every four years, that's 10 bites at the cherry. Therefore, the odds of becoming US President for any American born now stand at **1 in 13,500,000**. Statistically, it varies between men and women. The average life expectancy for women is higher, giving them a possible 11 bites at the cherry. That's a probability of **1 in 12,300,000**, while for men it's a less probable **1 in 15,000,000**. However, history has shown this not to be the case. There has never been a female US President. Margaret Thatcher became the

first (and, to date, only) woman to be elected Prime Minister of Great Britain. Not only did she overcome the odds of being a woman, she became the longest-serving Prime Minister of the 20th century, and the seventh longest ever, holding power for 11 years and 209 days.

Long shots

The idea of Nelson Mandela gaining power in South Africa was inconceivable in 1979, the year Thatcher came to power. He was in the 16th year of a life sentence, still incarcerated on Robben Island, and would serve a total of 27 years behind bars. Yet in the year Thatcher finally left office, Mandela walked out of prison and began a campaign of negotiations that, four years later, would see him become President of South Africa in the country's first-ever multi-racial election.

Mahatma Gandhi began his political career in South Africa and, like Mandela, served time in prison, both there and in India. As an Indian, albeit a well-educated one, trying to give his people dominion over their own land, he had to overcome huge odds, including bringing an end to 300 years of British rule. That he should die violently, shot by an assassin, shortly after gaining independence for India, did nothing to undermine what he achieved through non-violent protest and civil disobedience.

THEY WANNA PUT ME IN THE MOVIES

Who wouldn't want to be a movie star? Huge pay cheques, an adoring public, flunkies running around after you 24/7… and all you've got to do is act naturally.

But what are the chances of getting discovered? Thousands of hopefuls flock to Hollywood every year, yet only a handful get the lucky break. If you want to improve your odds of making it in the movies, you need to follow the basic rules:-

1. BE IN THE RIGHT PLACE AT THE RIGHT TIME

Olympic swimming champion Johnny Weissmuller was enjoying a dip at the Hollywood Athletic Club when MGM scriptwriter Cyril Hume happened to saunter by. Hume was writing the script for *Tarzan the Ape Man* and was on the lookout for someone to play lead. Six-foot-three Weissmuller was given the part and went on to make the role his own.

2. HAVE THE RIGHT PROFILE

French soccer player Eric Cantona made a name for himself in the 1990s as a player of temperamental genius. One minute he was stroking 25-yard volleys into the corner of the goal, the next he was kung-fu kicking a spectator who had insulted him. At the age of 30 he decided he fancied being a film star, and surprised everyone by landing a part in the movie *Elizabeth*, starring Cate Blanchett (Cantona played a bolshy Frenchman). Eleven years and as many films later, he was starring in the eponymous *Looking for Eric*, nominated for a Golden Palm at Cannes.

3. TRY, TRY AND TRY AGAIN

Two-time Oscar winner Dustin Hoffman has been a Hollywood hit for almost half a century, but he had to wait for success to come his way. Having embarked upon an acting career in 1956, it would take another 10 years before he landed the role that would make his name as a film star, as Ben Braddock in *The Graduate*. Hoffman was 30 when the film premiered. In the meantime he was building a stage career and made ends meet by taking a variety of jobs in restaurants, typing pools and factories, as well as teaching.

4. DON'T TAKE NO FOR AN ANSWER

Charlize Theron had no intention of becoming an actress when she went to Hollywood. The South African wanted to be a dancer. But when her bank refused to cash a cheque for her, an argument ensued that would make the amount on that cheque pale into insignificance. A talent spotter in the queue behind persuaded her to try her hand at acting and a string of box office hits quickly followed, including an Oscar-winning portrayal of Aileen Wuornos in the 2003 film *Monster*.

SO YOU WANT TO BE A PAPERBACK WRITER

They say that everyone has at least one book in them. But how many of us ever get round to writing that novel? And how many succeed in getting it published?

It's not unusual to have a book published, mind you. In the UK, which tops the UNESCO index of books produced in each country per year, there are more than 500 books published every day on average. Surely one of those could be yours.

Of course, some authors publish more than their fair share. Barbara Cartland, the queen of historical romance, was knocking out 20 a year towards the end of her prolific career.

There again, some books are written by more than one author. So you can assume an average of one author per book per year, and on that assumption there are approximately one million published authors each year. That's roughly **1 in every 6,700** people on earth.

But 16 per cent of the world's population are either too young to read

or have never been taught to do so. This improves your chances, assuming you are literate, to **1 in 5,600**. Put another way, for every town the size of St Tropez in the South of France, there should be one published author each year.

SHAKESPEARE AND THE CHIMP

The concept of a chimpanzee typing the works of Shakespeare is one that has been used to explore infinity. The argument goes that if you get enough chimps typing at random for enough time, one of them will produce the complete works of Shakespeare. It sounds ridiculous but it's true… in theory anyway.

If you had just one chimp typing letters at random and you wanted it to type the M of Macbeth, it would have a **1 in 26** chance of doing so (assuming you removed the non-letter keys from the equation). Its chances of then typing the A would be 1/26 x 1/26, which equals **1 in 676**. By the time you get to the H, the odds will have increased exponentially to **1 in 8,031,810,176** (26 to the power of 7).

However, if you had 8,031,810,176 chimps…

Without delving too much further into the mindblowing mathematics, the title of *Macbeth* features just seven of over 96,700 characters (including spaces) in the whole play. (Adding a space bar to your chimp's typewriter would lengthen the odds further still.) And the complete works of Shakespeare amount to 37 plays, 154 sonnets and five poems. Therefore, the odds of a chimp typing all that are effectively zero. And, more disturbingly, you're more likely to get the complete works of Dan Brown first.

CRIME

Crime

1 in 3 – falling victim to crime in Australia

1 in 60 – being assaulted in Scotland

1 in 3,000 – being murdered in South Africa

1 in 10 – owning a gun

1 in 2,000 – being shot dead in Colombia

1 in 58 – being burgled in Denmark

1 in 143 – being in prison in the USA

ARE YOU FEELING LUCKY?

Unless you're a criminal yourself, you will no doubt be concerned by crime figures wherever you happen to live. But those with the biggest cause for concern are the Australians, whose annual risk of becoming a victim of crime is just under **1 in 3**.

We're not talking minor crimes here: this was the figure recorded for 11 serious types of crime, including robbery, burglary, theft and sexual assault. New Zealanders don't fare much better, though it should be pointed out that these figures are, to some extent, a reflection on a country's efficiency in reporting and collating crime data, and are not necessarily connected to the fact that both countries were partly populated by shiploads of convicts from Britain – which coincidentally came third in the United Nations survey.

Different nations seem to specialize in different types of crime. The Scots, for example, top the table for assault, while the Italians go in for car theft. Mind you, with all those Ferraris, Lamborghinis, Maseratis etc, the temptation there is particularly strong. So let's take a look at some of the more serious categories of crime and see how likely you are to become a victim.

ASSAULT

In 2006, there were 84,692 reported assaults in Scotland. This looks like a small number in comparison with the 419,101 reported in England and Wales, or the 834,885 in the USA, yet with a population of just over five million, it meant that an alarming **1 in 60** Scots suffered a non-fatal, non-sexual physical attack.

The United States, by contrast, was fortunate to see a rate of **1 in 355** in the same year.

Europe appears to be the assault capital of the world, with six countries featuring in the top ten. This is likely to be due in part to the way these crimes are recorded, although a heavy drinking culture in many northern European countries will also be a factor.

Assault Top 10	
Scotland ....**1 in 60**	Israel.................................**1 in 155**
Sweden................................**1 in 109**	Germany............................**1 in 159**
England and Wales..............**1 in 130**	Chile ....**1 in 174**
Belgium ....**1 in 144**	Maldives ....**1 in 189**
Finland................................**1 in 153**	Lesotho ....**1 in 206**
	Figures from UNODC

THICK AS THIEVES

Here's an interesting statistic. In 2008 there were 2,222,200 recorded burglaries in the USA. One of them took place at the Fred Meyer department store near Seattle, Washington, on 9 July. Police followed a trail of packaging and goods until they came across the perpetrators, asleep on their plunder of pillows, sleeping mats and a hammock! They even had time to photograph the culprits before waking them up and arresting them.

In another incident that same month, a burglar who had gained entry to the balcony of a Tampa woman's apartment was spotted walking down the road wearing a pair of tight blue shorts, later identified by the victim as hers! He was arrested and charged with burglary, petty theft, trespass and criminal mischief.

Peak time for crime

It's not surprising both incidents took place in July. A report from insurance company Aviva shows burglaries rise significantly in summer, peaking in August at 20 per cent above the monthly average. You're 28 per cent more likely to be burgled on a Friday than on a Sunday. Annual celebrations are also popular among burglars. In England, for example, burglaries rise by 25 per cent on November 5th, Guy Fawkes Night, when everyone is out enjoying fireworks displays – the perfect cover for a break-in.

Although it suffers the most burglaries year on year, the USA is not the place where you're most likely to be burgled. Denmark presents the greatest risk, followed by New Zealand and Australia.

Burglary Top 10

Denmark	1 in 58	Iceland	1 in 116
New Zealand	1 in 71	Belgium	1 in 118
Australia	1 in 87	Slovenia	1 in 135
England and Wales	1 in 94	Switzerland	1 in 135
Sweden	1 in 98	United States	1 in 140

Burglary is defined by the UNODC as 'gaining unauthorised access to a part of a building/dwelling or other premises; including by use of force; with the intent to steal goods'. Robbery, on the other hand, is theft from a person, 'overcoming resistance by force or threat of force'. You might think that robbery, therefore, attracts a different type of criminal, and the national statistics support that. According to the same UN data, only three countries feature in the top 10 for both burglary

and robbery. The likelihood of being robbed is also considerably less than that of being burgled, except in the worst countries in each case, Denmark and Belgium, where the risk is remarkably similar.

Robbery Top 10

Belgium	1 in 54	Russian Federation	1 in 580
Spain	1 in 94	France	1 in 582
Chile	1 in 219	England and Wales	1 in 679
Maldives	1 in 511	United States	1 in 705
Portugal	1 in 512	Italy	1 in 822

HEY! THAT'S MY CAR!

For most of us, other than our house (which few of us really own anyway), our car is our most valuable possession. We clean it, we polish it, we love it and we do everything in our power to protect it. Yet still, every day, thousands of us go to look for our cars and find them gone – the car thief has struck again.

Car theft is a multi-billion dollar business with around two million private cars being stolen in Europe and the USA alone each year – that's one every 15 seconds. The good news is that the numbers have been falling almost everywhere in recent years – with the exception of Italy.

The fact that car theft is so prevalent in Italy is no doubt due to the number of luxury sports cars in the country. While car thieves may have reduced the quantity of cars they steal, they now focus on quality, with luxury cars the prime target. According to law enforcement agency Europol, a thief who steals a high-end car could make €15,000 (£13,000/$20,000). And Europol says that €6.75billion (£5.8billion/$9.2billion) worth of cars are stolen and trafficked between European countries each year.

Private Car Theft Top 10			
Italy	1 in 323	Norway	1 in 687
Sweden	1 in 474	Bahrain	1 in 754
USA	1 in 504	Belgium	1 in 766
Canada	1 in 507	Spain	1 in 1,047
Czech Rep	1 in 600	Latvia	1 in 1,209

If you include motorcycles and other motorized vehicles, Italy drops down the list to fourth place, behind Bermuda, France and Sweden. There were 858 cars stolen in Bermuda – **1 for every 74** residents. This may have something to do with the heavy restrictions on car ownership on the island (one per household) and the fact that visitors to the island are not allowed to rent cars or borrow private cars. In France, the chances of having a motor vehicle stolen are just above **1 in 100**. Indeed, it is often not for the car as a whole but for parts that the theft occurs. In America, the National Insurance Crime Bureau's list of the most frequently stolen cars of 2008 was topped by the rather unexciting but hugely popular 1994 Honda Accord, followed by the 1995 Honda Civic and then, incredibly, the 1989 Toyota Camry!

STRANGER DANGER

The greatest fear for any parent must be losing a child – or rather, having a child stolen from you. Cases such as that of Austrian Natascha Kampusch, who was snatched at the age of ten by Wolfgang Priklopil and held captive in his cellar for eight years, or of British three-year-old Madeleine McCann, who went missing from her bed while on a family holiday in Portugal in 2007 and is still missing, fill parents with dread and make them draw their own children closer to them.

The publicity surrounding abductions such as these, and worse still cases where children have been taken and killed, has changed the way we protect our children. A survey in Britain found that the percentage of children who make their way to school unaccompanied

by an adult had fallen from 22 per cent in 2002 to just 14 per cent in 2008. And 29 per cent of parents cited fear of assault or molestation as their reason for accompanying their children to school.

But child abduction is nothing new. One of the most famous cases in history involved Charles Lindbergh, the first man to fly non-stop across the Atlantic. On 1st March, 1932, his 20-month-old son Charles Jnr was kidnapped from the family home in New Jersey and a ransom note left behind. Two months later, while investigators struggled in their efforts to track down the culprit, the child's body was discovered in undergrowth not far from the Lindbergh's home, its skull fractured. It took a further two years for the police to get their man, Bruno Hauptmann, who was found guilty and sentenced to death by electric chair. Hauptmann protested his innocence to the end. As a result of the Lindbergh case, kidnapping was made a federal offence in the USA. And child abduction became one of the favourite stories for headline-hungry newshounds.

According to UK Home Office figures, nine per cent of all reported abduction offences involved a child being kidnapped by a stranger – a probability of **1 in 218,000**. Just over half of the victims were girls and the average age was 10. Five times as many abductions were perpetrated by one of the child's parents, or by people known to the child.

So while there is a very real danger from strangers that children should be warned about, the odds are far greater that a child will be taken by somebody from within its own family.

MURDER

Have you ever fancied a trip to Latin America but been put off by its violent reputation? How about Africa? There's so much to see in these lively and passionate continents, but is it worth the risk? You hear so many horror stories, but are they just scaremongering?

The chilling fact is that 29 of the 30 worst countries for intentional homicide are in Africa or Latin America and the Caribbean.

And the murder rates are frighteningly high. People think of the USA as being a violent country because of the number of cop shows that come out of American TV studios, but the USA's intentional homicide rate of **6 in every 100,000 people** is less than 10 per cent of the rate in South Africa, which tops the table, and El Salvador, which is second. The latest UN statistics on homicide around the world show more than **1 in 1,500** South Africans being murdered each year*.

By comparison, the rest of the world is remarkably safe. It's not until you get to 28th in the table that the first Asian country appears (the Philippines, at **1 in 4,800**), while Russia is the most dangerous European country, with a murder rate of **1 in 4,950**.

The rest of Europe falls well down the list, with Estonia and Ukraine recording a murder rate of **1 in 11,000**. In most of western Europe the chances of being a victim of intentional homicide go out to around **1 in 100,000**.

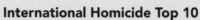

International Homicide Top 10

South Africa	**1 in 1,500**	Democratic Republic	
El Salvador	**1 in 1,600**	of the Congo	**1 in 2,900**
Colombia	**1 in 1,900**	Zimbabwe	**1 in 2,900**
Ivory Coast	**1 in 2,000**	FIGURES ROUNDED TO THE NEAREST 100.	
Guatemala	**1 in 2,400**	[*In 2010, the South African police	
Angola	**1 in 2,600**	claimed that the murder rate had	
Burundi	**1 in 2,700**	fallen by 8.6 per cent, but still stood	
Sierra Leone	**1 in 2,700**	at more than **1 in 3,000**.]	

Guns play a huge part in these statistics. According to the International Action Network on Small Arms (IANSA), 1,000 people die every day from gunshot wounds. A quarter of these occur in wars, the rest are homicides, suicides and accidents. Ninety per cent of victims are young men and boys. IANSA puts the number of firearms in the world today at 875 million, 74 per cent of which belong to civilians. That means roughly **1 in 10** people on Earth own a gun.

Fatal Shootings Top 10

Colombia...........................**1 in 2,000**	Venezuela ....**1 in 4,800**
Honduras..........................**1 in 3,300**	Guatemala..........................**1 in 5,800**
El Salvador ....**1 in 4,500**	Jamaica...............................**1 in 5,900**
South Africa.....................**1 in 4,500**	Ecuador ....**1 in 9,800**
Brazil................................**1 in 4,700**	Philippines.......................**1 in 10,400**

FIGURES ROUNDED TO THE NEAREST 100

CALL THE COPS!

There's never a policeman about when you want one, or so the saying goes. And yet the minute you creep over the speed limit, or try to nip down the bus lane, there's one right there flagging you down.

London's Metropolitan Police Force was the archetype for most modern police forces around the world. When it was formed in the early 19th Century, it consisted of 17 divisions of 144 constables: a total of 2,448 bobbies (they took their nickname from Robert Peel, the British Home Secretary who implemented the Metropolitan Police Act of 1829). With the population of London around two million at the time and rising fast, that meant one constable for every 800 people. Given that they weren't all on duty at once, and that those who were were frequently drunk, the chances of getting prompt and effective police assistance were slim.

Today the Metropolitan Police employs approximately one constable for every 200 people, and with London now covering an area of 609 square miles, that would be one for every 52,000 square yards if they were all on duty at the same time – which they're not.

In Britain as a whole, the ratio of police officers to population is **1 to 500**, the same as France, Australia,

Switzerland, Netherlands and Denmark. But do more police officers mean less crime? Apparently not.

If you want the reassurance of being guaranteed a policeman close at hand, move to Montserrat. This volcanic island in the Caribbean boasts one police officer for every 128 inhabitants, making it the

most policed country in the world. Yet Montserrat lies seventh in the table for total crimes per capita, just below the UK and just above the USA, with two crimes committed for every 25 inhabitants each year.

In Dominica it's even worse. There, one crime is committed for every nine inhabitants, despite the presence of one police officer for every 156 people. Like Montserrat, Dominica is part of the Lesser Antilles, an archipelago of small volcanic islands.

Police Presence around the world

Montserrat	1 in 128	Australia	1 in 500
Italy	1 in 180	United Kingdom	1 in 500
Germany	1 in 345	Japan	1 in 555
South Africa	1 in 360	Canada	1 in 560
United States	1 in 445	India	1 in 1,000

CRIME AND PUNISHMENT

Let's put the boot on the other foot. Say you are a criminal; you might want to know your chances of getting caught and punished. According to the UN data, approximately one third of all crimes committed lead to an arrest, of which 58 per cent result in a conviction. In other words, only a fifth of all crimes get punished.

If the idea of going to prison worries you, don't commit your crimes in the USA. There the penal system is particularly keen on locking criminals up, with 0.7 per cent of the population currently serving time behind bars.

Incarcerated Top 10			
USA	1 in 143	Lithuania	1 in 313
Belarus	1 in 194	Moldova, Republic of	1 in 390
South Africa	1 in 249	Chile	1 in 426
Latvia	1 in 276	Azerbaijan	1 in 427
Panama	1 in 284	Romania	1 in 464

The ultimate punishment, the death penalty, is still imposed in more than 50 countries. According to Amnesty International, at least 2,390 people were executed in 2008, a figure that only represents documented executions and is, therefore, likely to be considerably higher in reality.

The five nations that carried out the most executions, China, Iran, Saudi Arabia, Pakistan and the USA, accounted for 93 per cent of all executions.

So where in the world are you most likely to find yourself faced with

the death penalty? Of the countries that consistently apply the death penalty, Iran presents the biggest fear for criminals hoping to escape with their lives. **1 in 210,000** Iranians were executed in 2008, slightly up on the number in 2007. And some of those were stoned to death.

In neighbouring Iraq, two years after Saddam Hussein fully paid his dues, the death sentence was carried out on **1 in 850,000** people.

In the USA there were 37 executions in 2008, roughly **1 in 8 million**, and 18 of them took place in the state of Texas. The next highest was Virginia with four. But the total of 37 was the lowest since 1994, continuing a trend that has seen America move steadily away from use of the death penalty. In Asia, however, the mood is not so merciful. Asia accounted for more executions in 2008 than the rest of the world put together, due in no small part to China carrying out the death sentence on at least 1,718 people – **1 in 770,000**.

Executions Top 5		
Iran ....**1 in 210,000**	Pakistan .... **1 in 4,690,000**	
Saudi Arabia.................**1 in 270,000**	United States .... **1 in 8,160,000**	
China ....**1 in 770,000**		

HEALTH

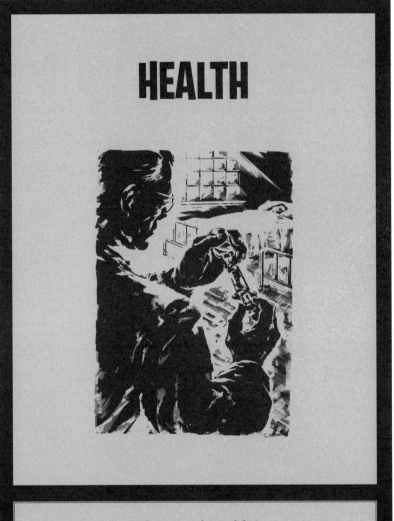

What are the odds?

Evens – living beyond the age of 50 a hundred years ago

Evens – living beyond the age of 32 in Swaziland today

1 in 4.25 – a newborn girl in the UK living to 100

1 in 200 – having the HIV virus

1 in 500 – being diagnosed with cancer in any given year

1 in 900 – dying of heart disease

1 in 10 – being obese by the year 2015

PRESERVING LIFE AND LIMB

If you should be taken ill and need to go into hospital, what are the chances of there being a bed available for you?

In the 10 years from 1998 to 2008, the number of hospital beds per 1,000 population fell from 6.3 to 5.5. This was among the members of the Organisation for Economic Co-operation and Development (OECD), more loosely known as the First World – in other words, the world's wealthier nations.

This average figure of 5.5 beds per 1,000 population (**1 bed per 182 people**) is fairly representative of most of the OECD countries, although the figures do range right down to 1.7 per 1,000 (**1 bed per 588 people**) in Mexico and right up to 13.8 per 1,000 (**1 bed per 72 people**) in Japan.

The fact that Japan has four-and-a-half times as many hospital beds per capita as the USA, and two-and-a-half times the OECD average, suggests that the country has a greater health problem than most. But this is not the case. The average life expectancy in Japan is higher than every other county on Earth, at 82.6 years.

The global average life expectancy is 67.2 years, which also happens to be the average life expectancy in Uzbekistan. Interestingly, it is also double what it was 200 years ago.

A LONG LIFE

Another birthday?

In the developed world, your chances of a long life are increasing. Taking the USA as an example, between 1950 and 2000 the life expectancy of a male at birth rose from 65.6 (below the current global average) to 73.7. Females are faring even better. Between 1950 and 2000 their life expectancy increased from 71.1 to 79.4.

A hundred years ago the picture was bleak, with an average life expectancy of 51.5. The average American male was not expected to live beyond the age of 50.

Today most of us stand a better chance of reaching 100! If life expectancy continues to increase at its current rate, there are expected to be over one million centenarians living in the USA by 2050.

In the UK, the number of centenarians has increased by a factor of 90 in the last 100 years. In 2007, British centenarian Alec Holden collected £25,000 ($40,000) from the bookies, having bet £100 ($160) 10 years earlier that he would live to celebrate his 100th birthday. William Hill offered him 250-1. In other words, they rated his chances of reaching 100 at just 0.4 per cent.

At the time of his 100th birthday, the chances of a newborn boy living to 100 were reckoned to be 18.1 per cent, and 23.5 per cent for a girl. That's nearly **1 in** 4! And the UK is by no means exceptional.

AIDS

Revealingly, 125 countries come above the global average for life expectancy, while 68 fall below it.

The sad fact is that while health issues such as heart disease and stroke have been rendered far less of a common threat to life in the First World, AIDS has taken a heavy toll in the Third World.

1 in 200 people on Earth were living with HIV at the end of 2008.

1 in 2,500 became infected with the HIV virus.

Two-thirds of new infections took place through heterosexual sex.

The threat of AIDS is most deadly to the young. Around half of new infections are in people under 25, and if you're aged between 20 and 24, AIDS is now the second most common cause of death.

1 in 3,300 people died from AIDS in 2008.

If you're surprised by these figures, you're probably not living in sub-Saharan Africa. This region alone accounts for almost three-quarters of the world's population with HIV.

Worst hit of all is Swaziland.

1 in 4 adults in Swaziland have the HIV virus.

And it is most prevalent among women.

1 in 3 women aged 15 to 49 have HIV.

For men within the same age group, the prevalence is **1 in 5**.

LIFE EXPECTANCY IN SWAZILAND IS JUST 32.

I WILL SURVIVE

According to an article published by Harvard Medical School in 2006, your chances of living to 100 increase significantly if you can first make it to 65. Of course, if you don't make it to 65, your chances of reaching 100 are zero, but that's not what they meant.

The point they were making is that anyone who lives to a healthy 65 years of age has beaten the odds of succumbing to most of the dangers that cause an early death, be it disease or violent action.

HEART DISEASE AND STROKE

These figures from the American Heart Association show how the risk of heart disease increases with age.

Age	Risk of coronary heart disease	
	Male	Female
20–39	1 in 140	1 in 140
40–59	1 in 14	1 in 14
60–79	1 in 4	1 in 6
80+	1 in 2.7	1 in 4.5

It's also interesting to note that the risk increases by a factor of 10 in the 40–59 age bracket.

1 IN 226 – the incidence of coronary attacks in the USA

Coronary heart disease (CHD) is the biggest single cause of death in the world. The chances of getting it are roughly equal for men and women up to the age of 60, but beyond that age men are at a far greater risk.

Diseases of the cardiovascular system claim 17 million lives each year – more than any other cause of death.

The cardiovascular system consists of the heart and blood vessels. The most deadly forms of cardiovascular disease are heart disease and stroke.

Cardiovascular disease is a narrowing of the blood vessels caused by a build-up of fatty deposits. This hinders the blood flow, and it's when the blockage becomes severe that a heart attack or stroke will occur.

Major factors that contribute to the build-up of fatty deposits are a poor diet, lack of exercise, consumption of tobacco and excessive consumption of alcohol. High blood pressure and high cholesterol are danger signs.

Each year:

1 in 900 people in the world die of heart disease

1 in 1,150 die of strokes.

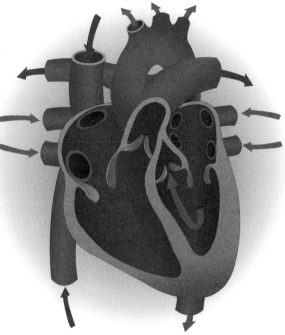

These figures from the American Heart Association show how the risk of stroke increases with age.

Age	Risk of stroke	
	Male	Female
20–39	1 in 330	1 in 160
40–59	1 in 100	1 in 37
60–79	1 in 13	1 in 13
80+	1 in 7	1 in 8

CANCER

1 IN 500

That's roughly how many new cases of cancer are diagnosed each year.

60 per cent of cancer deaths occur in the developing world.

In the developed world, your chances of surviving 10 years with cancer are about 45 per cent, though there is a marked difference between men and women.

For women it is 52 per cent.

For men it is 39 per cent.

According to the World Health Organisation, 7.9 million people died of cancer in 2007. That's more than **1 in 1,000**. And that figure is expected to rise to 11.5 million in 2030. An estimated 12.7 million new cases of cancer were diagnosed in 2008.

Dying for a cigarette

These rising figures are being put down to a rapid increase in the number of people who smoke. In the 20th century, the world's cigarette consumption rose from 50 billion to 5,500 billion. And smoking is responsible for nearly a quarter of all cancer deaths. While the First World has seen a steady decline in the number of people who smoke, the developing world has taken over as the hotbed of tobacco consumption.

Back in 1948, when statistics were first gathered on the subject, **4 in 5** adult British men were smokers. By 1970, this had fallen to just over **1 in 2**. And today the chances of an adult British male becoming a smoker are about **1 in 5** – a complete inversion of the figures from half a century before.

These figures are fairly typical of the West. But in Asia, the opposite trend has taken place. In China, Mongolia, Cambodia, South Korea, Laos and Indonesia, the odds of an adult male being a smoker are about

2 in 3. It's this prevalence in the Asian male population that is accounting for the global rise in cancer.

Women are still far less likely to become smokers than men. Fewer than **1 in 8** adult females worldwide smoke, while the global figure for adult males is about **1 in 3**.

OBESITY

THE ODDS OF AN AMERICAN BEING OVERWEIGHT
ARE 2 IN 3.

But America is not the fattest country in the world. That distinction goes to American Samoa, where fewer than **1 in 10** people are NOT overweight! This sad fact is put down to the introduction of an American-style diet high in fats and sugars, which contrasted dramatically with their natural diet.

The fattest countries on Earth

Here are the top 10 countries from a World Health Organization survey showing the percentage of population that is overweight.

1. American Samoa	**93.5%**	6. Bosnia-Herzegovina	**62.9%**
2. Kiribati (Pacific islands)	**81.5%**	7. New Zealand	**62.7%**
3. USA	**66.7%**	8. Israel	**61.9%**
4. Germany	**66.5%**	9. Croatia	**61.4%**
5. Egypt	**66.0%**	10. UK	**61.0%**

By 2015, one third of the world's population will be overweight, and more than one tenth will be obese. The WHO defines overweight and obesity as 'abnormal or excessive fat accumulation that may impair health'.

If you want to check whether you are obese, you can calculate your body mass index (BMI). Simply divide your weight in kg by the square of your height in metres. This is only a rough guide, but if your BMI is over 30 you can consider yourself obese. If it is over 25, you are overweight. Statistics show that your risk of contracting chronic disease increases progressively as your BMI rises above 21.

Being overweight causes:

75 per cent of heart disease

50 per cent of diabetes

The cause of obesity is simple and preventable. It is a combination of poor diet and lack of physical activity. The table opposite shows that the overweight problem is most prevalent in countries where diets with high sugar and salt content are most commonplace. And as these diets have become more available around the world, so the problem has spread.

By eating foods that are high in vitamins and minerals and low in sugars and fats, and taking at least 30 minutes of moderate physical exercise a day, you reduce your chances of being overweight to practically zero.

UNDER THE KNIFE

As recently as the 18th century, surgeons doubled as barbers, and their main duty was removing damaged limbs before infection set in. However, infection usually set in anyway.

Few patients survived any sort of surgical operation. Caesarean sections were carried out to save the child, not the mother. In most cases the mother was beyond saving. Today, the survival rate for a C-section is 99.98 per cent.

That's how far surgery has progressed.

These are averages across all age groups. Younger patients have a better chance of survival. Surgery to the vital organs is usually the treatment for a life-threatening condition which will often remain a threat after the surgery. Therefore, the diminishing survival rate over time is not a reflection on the success of the surgery, but on the severity of the condition.

Heart surgery

Survival rate: 98.4 per cent.

Kidney transplant

Survival to one year: 96 per cent

Survival to three years: 91 per cent

Liver transplant

Survival to one year: 85 per cent

Survival to five years: 65 per cent

Brain surgery

Survival to one year: 36 per cent

Survival to five years: 15 per cent

Survival to 10 years: 10 per cent

THE EYES HAVE IT

Lasik eye surgery has become increasingly popular for people with sight defects, and the process has been nearly perfected. Some operators claim a 98 per cent success rate in restoring 20/20 vision to myopic patients.

GENDER DYSTOPIA

The desire to change sex is estimated to be prevalent in **1 in 5,000** people over the age of 16. Of these, 80 per cent are male. Between 2 and 5 per cent of men habitually cross-dress as an expression of femininity, and studies have shown that a fifth of transvestites express the desire to change sex. Secrecy shrouds the numbers who actually go under the knife each year, but it's estimated to be anything from 1,000 to 2,500 per year worldwide.

NIPS AND TUCKS

Cosmetic surgery actually saw a decline in 2008. This was put down to the recession – an indication of the scale of importance of most cosmetic surgery. Even so, **1 in 175** Americans underwent some form of cosmetic surgery in that year alone.

Women made up 92 per cent of all cases. That means, even in a recession, the odds of an American woman having cosmetic surgery are **1 in 190**. And if she happens to be the wife of a plastic surgeon, no doubt they're even higher than that. Breast enlargement became the most popular operation, with an incredible **1 in 420** American women having their breasts enlarged in 2008 alone.

Breast reduction (gynecomastia) was the fourth most popular surgery among men, with **1 in 7,850** American men presenting themselves to have their moobs ironed out.

WHICH LEG WAS IT AGAIN?

We all know that there are risks involved in surgical operations, but we content ourselves with the conviction that the surgeons into whose hands we place our lives are highly professional, methodical and scrupulously thorough.

However, in 2009 a Dr Atul Gawande published a study that suggested the death rate from operations could be reduced by 40 per cent if doctors and surgeons used a simple flight-check style checklist. Why? Because every year patients emerge from theatre minus perfectly healthy body parts that have been removed by mistake.

The shocking stories include numerous cases of the wrong arm or leg being amputated, the wrong disc being removed in spinal operations and the wrong hip being replaced. One man had the wrong testicle removed and a woman was given a hysterectomy that she didn't need after her records were mixed up with those of another patient.

It's hard to imagine how such mistakes can occur, but they do, and with surprising regularity. In one incident a doctor got the addresses of his patients mixed up and circumcised the wrong child!

Any part of the body that can be removed or transplanted,

it seems, has been the subject of at least one such error. Arms, legs, eyes and internal organs are removed in error every year, and once they're gone, there's no putting them back. Known as 'wrong-site surgery', it occurs in an estimated **1 in 70,000** cases.

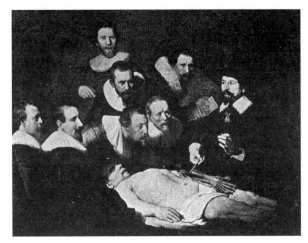

A report in America estimated that as many as 3 per cent of hospital patients are victims of hospital errors and 34,000 patients die every year as a result.

Teeth are the most commonly mistaken body parts when it comes to surgical removal, which suggests that size is a factor. If that's the case, cover your eyes – literally. A survey in New York found that 69 out of every one million eye operations resulted in 'surgical confusions'.

People with tattoos should beware. It is common practice for surgeons to mark the limb or whatever is to be removed by writing their initials on it in pen. There have been cases where a tattoo was mistaken for the surgeon's signature and the wrong eye was extracted.

ANYONE SEEN MY SCALPEL?

Have you ever come out from an operation and had the feeling that there's something cold and metallic rattling around inside you?

If so, you were probably one of the estimated **1 in 1,000-1,500** victims of what medical reports call 'retention of a foreign body'. Or, in layman's terms, 'the surgeon leaving his tools behind'.

HISTORY

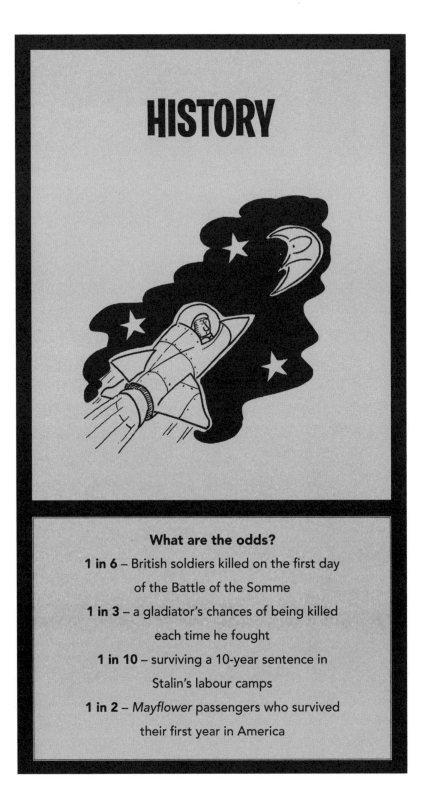

What are the odds?

1 in 6 – British soldiers killed on the first day
of the Battle of the Somme

1 in 3 – a gladiator's chances of being killed
each time he fought

1 in 10 – surviving a 10-year sentence in
Stalin's labour camps

1 in 2 – *Mayflower* passengers who survived
their first year in America

PRO PATRIA MORI

THE SOMME

The Battle of the Somme, which raged over five months from July to November in 1916, is considered one of the most scandalous wastes of life in the history of war. By the time the fighting ended, nearly 1.5 million men had been killed.

That the battle should have dragged on so long is remarkable, given the scale of the carnage on the very first day. 1 July 1916 has gone down as the most disastrous day in British military history. But of the estimated 120,000 men who went 'over the top' that day, very few had any notion of the appalling odds that confronted them.

General Haig had begun the campaign with a bombardment of the German lines, intended to break through the barbed wire and take out their fortified positions. After this, the infantry advance was expected to be a mere formality. The men, mostly first-time volunteers, who had been rapidly drafted to bolster the British Army after the first two costly years of the war, were ordered to walk, not run, in lines abreast, five yards apart, carrying 70lb of equipment – roughly half their bodyweight.

In effect, they were instructed to present themselves like targets at a fairground stall. And the German machine gunners, who had been waiting seven days for their moment to retaliate, took the invitation with ruthless efficiency. In that single day, the British Army lost 20,000 men. Only **2 in 5** officers survived.

SITTING DUCKS

The machine gun was the chief reason for the mass slaughter in World War I. Because of this, machine gunners became the primary target for advancing troops. By the Second World War, it was more dangerous to be sitting at the trigger of a machine gun than to be advancing towards it. As the war in the Pacific reached its most intense, the life expectancy of a US Marine machine gunner in the heat of battle was reckoned to be less than 20 seconds.

STALINGRAD

Similarly poor odds faced the Russian Army at Stalingrad in World War II. The average life expectancy of a reinforcement soldier was 24 hours. Of the 10,000 soldiers who made up the elite 13th Division, only three per cent survived, one third of them having been killed in their first day of fighting.

RORKE'S DRIFT

Other famous battles are remarkable for their low casualty rate. At Rorke's Drift in 1879, in which 139 British soldiers fended off several thousand Zulus, only 17 Brits lost their lives. Of those who survived, **1 in 11** were awarded the Victoria Cross – for valour in the face of the enemy – the highest number ever awarded to one regiment for a single action.

BATTLE OF YORKTOWN

The Battle of Yorktown, which marked the end of the American War of Independence, saw the US forces triumph with very few casualties. Of the 8,800 American troops, only **1 in 110** lost their lives. Their French allies sustained heavier losses, at a rate of **1 in 39**, while the British paid the heaviest price, losing 500 of their 6,000 men, a rate of **1 in 12**. Even so, the odds of a British soldier surviving this battle were twice as good as those soldiers who walked out on the first day of the Somme.

VIETNAM

In 1968, at the height of the conflict in Vietnam, the Americans and their allies from South Vietnam, South Korea, Australia and New Zealand were losing men at a rate of **1 in 33**. Since Vietnam, the odds of surviving wounds sustained in battle have improved greatly. During that conflict, **1 in 3** soldiers died as a result of their injuries, whereas **1 in 8** soldiers wounded in the Iraq war died.

STALIN'S TERROR

Nobody can say for certain how many people Joseph Stalin was responsible for killing, either through his brutal policies (his enforced famine in 1932-33 is reckoned to have claimed 6–7 million lives), or through the Great Purges of the late 1930s.

During this period, Stalin set about establishing an unshakeable grip on the Soviet seat of power by inflicting terror and paranoia upon people of all walks of life, from peasants to leading members of the Communist Party. A popular figure for the number of people killed under Stalin is 20 million. If that's the case, it means that the chances of falling victim to Stalin's murderous dictatorship were roughly **1 in 8**.

Those closest to Stalin faced the worst odds. At the 17th Party Congress in 1934, Stalin found himself facing a significant movement to remove him as General Secretary and replace him with his friend Sergei Kirov. Not long after, Kirov was assassinated. So began the Great Purges. In September 1936, 16 party members were put on trial for the murder of Kirov. All 16 confessed and were executed. The confessions had been written for them. It was the first of three show trials that took place between 1936 and 1938, which resulted in a total of 54 senior party members being executed or sent to the labour camps. The survival rate at the camps was **1 in 10**.

That fateful Congress in 1934 was attended by 1,966 party members. Little did they know that their chances of surviving the reign of terror over the next five years were less than **1 in 2**. Such was the thoroughness of the NKVD, Stalin's security service, in purging the Soviet Union of 'enemies of the people'. They also executed around 30,000 officers of the Red Army, including 90 per cent of its senior leadership, all of which left the Soviet Union bereft at the onset of WWII, leading to the further loss of millions of lives.

The NKVD was led by the ruthless Nikolai Yezhov, but he too fell foul of Stalin's terrible plan. Relieved of his post in November 1938, Yezhov was denounced by his successor, Lavrenty Beria, in March the following year, arrested, forced to confess to a catalogue of crimes, and executed in February 1940. Yezhov wasn't alone. Of all the NKVD officials working under him at the start of the purges, only 5 per cent survived.

THE MAYFLOWER

They sailed in search of freedom. But few survived the winter.

When the 'Pilgrim Fathers', their families, servants and hired hands, set sail from Plymouth, England, on 6 September 1620, they knew they were embarking on a journey into the unknown. But it was either that or stay behind and endure a life of persecution for their religious beliefs. The pioneers who established the first permanent European colony in America firmly believed it was a risk worth taking. They may have thought differently had they known that they only had a 50-50 chance of surviving the year.

The warnings were there. Of the 500 who had endeavoured to settle at Jamestown, Virginia, a decade earlier, only 61 had survived.

By the time the *Mayflower* arrived at Cape Cod, the pilgrims had suffered their first loss, a servant boy named William Button, as well as a member of the crew. Five more would die while still aboard in harbour. By spring, **45 of the 102** who had sailed (not including the crew) had perished, mostly as a result of scurvy, pneumonia and tuberculosis.

By the anniversary of their landing at Plymouth, New Jersey, only 53 remained – just over half of their original number.

Worst hit were the women. Though all 18 survived the journey, 13 failed to last the winter. A 14th died in June.

Yet against these odds, the colony kept going and became the basis for British colonization of North America.

THE FINAL FRONTIER

ONE SMALL STEP FOR MAN, ONE GIANT WIN FOR DAVID THRELFALL

'We choose to go to the moon. We choose to go to the moon in this decade and do the other things, not because they are easy, but because they are hard…'

On 25 May 1961, President Kennedy gave a speech at Rice University, Houston, Texas, in which he announced America's ambition to put a man on the moon by the end of the decade. Despite the fact that the Soviets had just put the first man into space (Yuri Gagarin) and had already crash-landed a craft on the moon two years earlier, many people thought this was a ridiculous statement by the President. The odds were stacked overwhelmingly against a successful moon landing.

But in Preston, England, a young man named David Threlfall took the President's words to heart. He watched the space race unfold with interest, and in 1964 he wrote to bookmaker William Hill. 'I'd like to bet £10 ($16) that a man will set foot on the surface of the moon before the first of January 1970.'

After some discussion, Hill's offered him odds of 1,000/1. Threlfall duly placed his bet on 10 April 1964 and the rest is history.

News of his wager spread rapidly around the world and thousands followed suit, though the odds soon plummeted. As the Soviets and Americans succeeded in making several unmanned soft landings on the moon, it looked more and more likely that Kennedy's forecast – and Threlfall's gamble – had not been so far-fetched after all.

When Neil Armstrong took his first step on the moon on 21 July 1969, 17 months within William Hill's own revised deadline, Threlfall was presented with his cheque for £10,000 ($16,000) live on television.

He bought himself an E-type Jaguar with his winnings, but tragically went on to kill himself in that very car.

Typically, the bookies survived and today William Hill is still taking bets from conspiracy theorists, who believe that the moon landings never happened. At odds of 100-1, you can bet that Armstrong's moment in history will eventually be proved to have been a hoax.

Meanwhile, more and more people are placing bets that intelligent life will be found to exist elsewhere in the universe. The USA are favourites, at 2-1 on, to make first contact, followed by Russia, China, France, UK and India.

LIFE AND DEATH IN THE ARENA

In terms of mass entertainment, the gladiatorial arena could be viewed as the precursor to televised sport. For 668 years, between 264BC and AD404, gladiators did battle for the amusement of the public, in hundreds of arenas scattered throughout the Roman Empire. The games were the people's pastime, and gladiators could become wealthy heroes, just like the professional athletes of today.

The big difference was that a gladiator went into each contest knowing he had about a **1 in 3** chance of being killed.

Most gladiators bit the dust in the first year of their career. If they could survive for three years, they were granted their freedom and the chance to live a feted life. But these were the lucky few. The Romans were disinclined to be merciful, and the remains found in special gladiator cemeteries in York, England and Ephesus, Turkey show that they died in all manner of gruesome ways.

In hand-to-hand combat, if they were not slain by their opponent then they fell to the mercy of the crowd. A gladiator who had not fought with vigour and panache would be dispatched by a sword through the throat and into the heart. Others were decapitated. Any thoughts of playing dead were foolish. At the end of the fight two characters would come into the arena, one with a red hot poker, the other with a mallet. If an apparently lifeless body

reacted to being touched with the poker, it would be finished off with a blow from the mallet.

Tooth marks from large carnivores in the remains found in the gladiator cemeteries stand as evidence of a *bestiarus*, a gladiator who fought against wild animals. Not content with watching men chop each other to bits, the Romans gloried in the slaughter of animals, and the

slaughter of people by animals.

Lions, tigers, bears, crocodiles, wolves, buffalo and other fierce creatures were used and abused in the name of entertainment. Criminals and Christians who were thrown to the lions stood no chance. They went in unarmed and were torn to pieces. But the *bestiarii* did have

a chance. Indeed, they often won. At the inauguration of the Colosseum in Rome, under Emperor Titus, 5,000 animals were put to death. By the time the gladiatorial games were abolished by Emperor Honorius in AD404, they had brought about the extinction of a number of species, including the North African elephant, the Caspian tiger and the Nile hippopotamus.

So who would win in a fight between man and beast? Let's look at the strengths and weaknesses of a selection of animals that a gladiator might have had to face in the arena.

GLADIATOR

| Height: **6ft** (183cm) | Bite force: **127psi** |
| Weight: **200lb** (91kg) | Speed: **22mph** (35km/h) |

Let's assume our gladiator is a superb physical specimen, with the build of a heavyweight boxer and the speed of an Olympic sprint champion. However, he is unarmed.

BROWN BEAR

| Height: **8ft** (244cm) | Bite force: **1800psi** |
| Weight: **700lb** (318kg) | Speed: **30mph** (42km/h) |

Standing on its hind legs, the brown bear dwarfs the gladiator by 2ft (61cm), giving it superior reach for those 4in (10cm) claws. The gladiator's only hope is to avoid close contact until the bear tires. But running away isn't an option. Neither is climbing a tree. The bear will be straight up there after him. And once it gets hold of him, it's no contest. Those powerful jaws could crush a man's head.

AFRICAN LION

| Height: **3ft at the shoulder** (91cm) | Bite force: **690psi** |
| Weight: **400lb** (182kg) | Speed: **36mph** (58km/h) |

Twice as quick as an Olympic sprinter, the lion makes up for its lack of height with powerful spring and hooked claws that grip as well as slash. Its 3in (7.5cm) canine teeth go straight for the jugular and don't let go until the fight is won.

TIGER

Height: **3ft 6in at the shoulder** (107cm)
Weight: **500lb** (227kg)
Bite force: **1,000psi**
Speed: **35mph** (56km/h)

The tiger is effectively a souped-up lion. Bigger, faster and more powerful, it does everything a lion can do only more so. With 4in (10cm) claws and canine teeth to match, once a tiger gets hold of our gladiator he'd wish he was fighting the bloke with the trident and the net instead.

BUFFALO

Height: **5ft 6in** (168cm)
Weight: **2,500lb** (1136kg)
Horn span: **40in** (101cm)
Speed: **35mph** (56km/h)

Weighing ten times as much as the gladiator, and able to run at nearly one and a half times the speed, the buffalo is practically unstoppable in full flight, unless you're a dab hand at the old Minoan trick of grabbing it by the horns and somersaulting over its back. Being behind the horns is definitely preferable to being in front of them. Not for nothing is the buffalo one of the 'big five' most deadly animals in Africa.

NILE CROCODILE

Height: **3ft** (91cm)
Weight: **1,500lb** (682kg)
Bite force: **2500psi**
Speed: **10mph** (17km/h)

Crocs may do most of their dirty work in the water, but they're a serious threat on land. They will chase you, so you'd better have your sprinting shoes on. The immense biting power behind those 60-odd teeth means that, in the words of Neil Diamond, when they know they have you, then they really have you. Your best bet is to force the jaws shut before they get hold of you as the muscles that open them are a bit puny.

MAN V BEAST

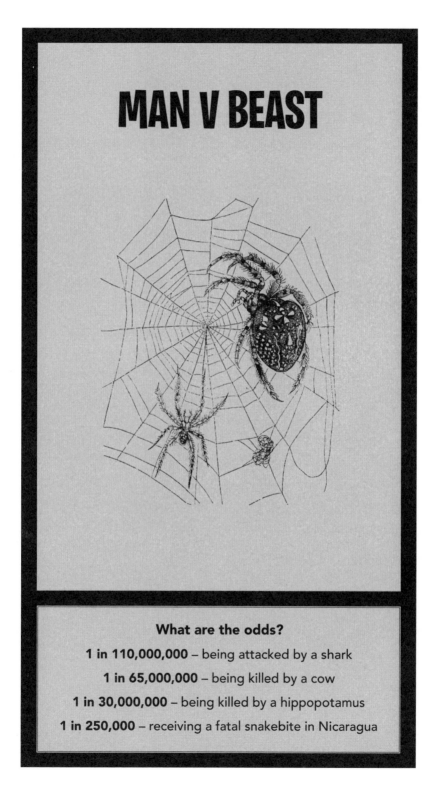

What are the odds?

1 in 110,000,000 – being attacked by a shark

1 in 65,000,000 – being killed by a cow

1 in 30,000,000 – being killed by a hippopotamus

1 in 250,000 – receiving a fatal snakebite in Nicaragua

♠ AN UNTIMELY END

There are many ways to die, and knowing the odds might just help you to avoid one or two of them. For example, if you know you stand a better chance of being swept away by a tsunami in South-East Asia than you do in northern Europe, you might just choose to settle in Denmark. However, before you make any such drastic life decisions, bear in mind that you're more likely to die of heart disease, cancer, stroke or respiratory ailments than to meet your end in the sort of dramatic style that makes the daily news.

America is an interesting case study. There, you can easily be involved in an air or road traffic accident, as well as finding yourself victim of a hurricane or earthquake. America also has its share of deadly animals – and people! A study by the National Safety Council worked out the odds of a wide range of causes of death.

Over a lifetime, here are the odds of an American being killed by:			
Heart disease	1/5	Plane crash	1/20,000
Road accident	1/100	Flood	1/30,000
Suicide	1/121	Earthquake	1/132,000
Shooting	1/325	Asteroid	1/500,000
Drowning	1/9,000	Falling coconut	1/650,000

Animal attacks are one of the most frightening causes of death, but the ones we're most afraid of actually pose the least threat. You've no doubt heard, for example, that the most deadly animal in Africa is not the lion or the crocodile but the hippopotamus. Yes, that lumbering herbivore is

Your chances of being killed by an animal:

Mosquito	1/100	**Hippopotamus**	1/650,000
Snake	1/1,500	**Jellyfish**	1/1,000,000
Bees	1/50,000	**Cows**	1/1,000,000
Scorpion	1/200,000	**Crocodile**	1/1,160,000
Tiger	1/200,000	**Dog**	1/5,000,000
Elephant	1/300,000	**Shark**	1/10,000,000
Lion	1/300,000	**Bear**	1/14,000,000

indeed a ferocious beast, prone to attacking humans and gouging lumps out of them with its massive teeth. And it can 'lumber' at 30mph, trampling and gnashing at anything in its path. One of the most horrific recorded attacks occurred in the 1970s, when a hunter named Spencer Tyron had his head and shoulders bitten clean off near Lake Rukwa in Tanzania by a bull hippo that had capsized his canoe. Hippos account for around 150 deaths each year, more than any of the other big animals in Africa, but the fact that hippos are only found in a relatively small part of the world means that if you do wander far and wide you have more chance of being killed by an elephant or a crocodile.

Sharks are probably the biggest fear for most humans, since they tend to attack out of the blue just when we're enjoying ourselves. But with only around 10 fatal attacks each year, most of us can go in the water fairly confident that it won't be us. In fact, we have ten times more chance of being stung to death by a jellyfish. Size isn't everything when it comes to deadly animals – by far the deadliest of all is the tiny mosquito, responsible for between one and three million deaths each year through the spread of diseases like malaria and dengue fever.

THE BIRDS

Alfred Hitchcock captured the full horror of being attacked by animals in his famous thriller *The Birds*. The story, from the novel by Daphne DuMaurier, focused on a California heiress, Melanie Daniels, who follows her boyfriend to a quiet seaside town. She receives a rather chilly welcome from the people there, and from the local birds, who seem to want to peck at her for no apparent reason. As the human hostility intensifies, so do the bird attacks, with deadly intent.

It's a classic piece of Hitchcock horror, but what are the chances of it happening in real life?

Well, in 2010, in the quiet English seaside resort of Peacehaven, East Sussex, the birds began attacking people in a frightening

re-enactment of the Hitchcock movie. Individual bird attacks are not uncommon, but in this case the attackers were swooping in gangs.

Displaying an abnormal level of aggression, seagulls started flying at people and pecking at them. A seagull's beak and claws can leave deep wounds, and if they get your eyes, you're in big trouble.

One victim, Rosemary Howat, was attacked by two seagulls, which had built a nest in her garden. 'It's got to the point where I can't go outside in case they come at me,' she said.

Conservationists said the birds had become aggressive after seeing their nests cleared from roofs and their eggs smashed in the process. Some birds had even been shot at – an offence in England that carries a £20,000 ($32,000) fine.

The biggest threat from birds to man comes when they fly into the engines of jet planes: bird strikes, as they are known

The other major threat from birds is avian flu, or the H5N1 virus, which, if it mutates to a form that can be transmitted from human to human, is feared to become as lethal as the flu pandemic of 1918. The Spanish Flu, as it became known, lasted until 1920 and infected **1 in 3** of the entire world population, killing **1 in 10** of those it infected.

Estimates of the severity of a new pandemic, should it happen, range wildly from 1 million deaths to over 100 million. The UN health agency went with a figure of 7.4 million – or **1 in 900**.

DON'T GO IN THE WATER!

That was a tagline for the movie *Jaws*, a film that caused more terror on beaches throughout the world than any movie before or since. The idea of a man-eating shark picking off victims at random in a popular resort struck horror into film-goers in 1975 – and continues to do so today. For *Jaws* was based firmly on fact. And anyone who goes in the water in areas where these fascinating predators are known to swim considers the odds of falling victim to shark attack.

So what are those odds? In crude terms, about **1 in 110,000,000** each year. But that doesn't tell the story. If you live in Siberia, you can rule out any threat of being attacked by a shark. On the other hand, if you live in Volusia County, Florida, you stand a better chance than almost anyone else on earth.

The USA is the scene of two-thirds of the world's reported shark attacks. Three-quarters of those attacks take place in Florida, and two-thirds of Florida's attacks occur in Volusia County. Popular for its seaside resorts, Volusia reports around 15 shark attacks each year on average.

In general, your chances of surviving a shark attack are good. The global average is about one fatality per 15 attacks, but these statistics

vary wildly. In 2009, for example, Australia reported 20 attacks, of which none were fatal. South Africa, in contrast, saw four fatal attacks out of six.

The type of shark has a major bearing on your chances of survival. Great whites, tiger sharks and bull sharks are the three most deadly species. They account for the majority of reported shark attacks as they inhabit the waters where people like to swim and surf. Great whites have been held responsible in more than a third of unprovoked attacks, of which a quarter have been fatal.

In for the kill

Bull sharks are not as easy to identify as great whites and so they are reported in fewer cases – about **1 in 8**, though in reality this figure should be higher. However, of all the cases in which bull sharks have been cited, the kill ratio is similar to that of the great white, about **1 in 4**, which matches the tiger shark. Other fearsome names, such as the hammerhead, blacktip, mako and nurse shark are far more likely to bite without causing fatal injury.

Surfers are by far the most at risk, accounting for about 60 per cent of attacks. **1 in 4** attacks are on swimmers, **1 in 10** on divers and the rest on people climbing in and out of the w a t e r. Curators of the International Shark Attack File point out you're about 76 times more likely to be killed by lightning than by a shark. And here's a thought, next time you forgo the seaside in favour of a country walk: you're twice as likely to be killed by a cow as by a shark. So relax. Or take your holidays in Siberia and keep away from anything that moos.

CREEPY CRAWLIES

Spiders, snakes, lizards, scorpions, insects... a fear of animals that make the skin crawl is among the most common phobias known to man. But what is there to be afraid of? How likely are these little creatures to do us any serious harm?

Well, that all depends on where you live. If you're worried about being stung or bitten by a deadly creepy crawly, steer clear of Mexico, that's my advice. Mexico tops the charts for the number of deaths from contact with spiders and scorpions and has the fourth-highest number of deaths from snakes and lizards. In fact, the world's only two poisonous lizards live in Mexico: the gila monster and its cousin, the Mexican beaded lizard.

These two, however, are very unlikely to harm you, being slow-moving and not particularly aggressive unless provoked. Far more dangerous are Mexico's many snakes, including rattlesnakes, jumping vipers (they strike so hard they actually leave the ground) and the fearsome Fer de Lance, which can grow up to 8ft (244cm) in length and bites without provocation. The venom of all these snakes is highly toxic and will kill if not treated immediately.

The Black Widow spider is a native of Mexico, which, in the 2004 World Health Organisation survey, reported 14 deaths from spider bites – a rate of **1 in**

7.5 million, similar to Australia, which has even more deadly spiders such as the lethal Funnel Web and the Redback.

But in a country the size of Mexico, with a population in excess of 100 million, the death rate from contact with these creepy crawlies is much lower than in smaller countries, where venomous animals abound but the anti-venoms required to treat their bites and stings are less accessible. In the Dominican Republic, a country with a 10 million population, the rate of deaths from spider bites is **1 in 1.65 million** – four and a half times that of Mexico. Meanwhile, in Nicaragua and Panama, the chances of receiving a fatal snakebite are around **1 in 250,000** – six times the rate in Mexico.

Deadly scorpions are most prevalent in North Africa, where there are several species that will do you in if you're not in a position to get prompt treatment. Egypt reported 78 deaths in 2004, at a rate of **1 in 1.04 million**. However, Mexico topped the list with 84 deaths, though the likelihood was less at **1 in 1.25 million**.

It tends to be the young and the elderly who succumb to venomous

bites and stings, but there are plenty of creepy crawlies out there that will bring down a fully grown man if he does not receive prompt medical attention. So beware! Or go and live in Iceland where there are no venomous animals at all.

RATS

The rat is a creature that strikes fear into the hearts of millions. Remember Winston Smith in George Orwell's *1984*? When they put him in the dreaded Room 101, he found himself face to face with what he feared most – rats. While rat lovers will argue that these are actually clean-living and friendly animals that can make a sweet household pet, history has shown the rat to be a carrier of deadly diseases.

Perhaps our innate fear of rats stems from the Black Death, the devastating plague that swept through Europe between 1348 and 1350, carried by black rats (*rattus rattus*). If you think about Europe today, the subjects that occupy all its citizens are mundane issues like currency, agricultural policy and sport. For those two years in the mid-14th Century, the word on everybody's lips was 'plague'.

Depending on which part of Europe you were from, your chances of falling victim to the Black Death were anything from **1 in 4** up to **1 in 2**. The plague divided communities, as the fear of contagion drove the healthy to avoid the contaminated, even to the extent that parents would not tend their own sick and dying children. Hence the expression, 'avoid him like the plague'.

Death was everywhere, so much so that it became a laughing matter. People abandoned hope, and with it went social order. Bodies were buried in mass graves, or plague pits, and those who were privileged enough to get their own would have a skull engraved on the headstone, to show they were victims of the plague.

And though the Black Death fizzled out in 1350, the plague returned from time to time, carried by those dreaded black rats. Though we blame the black rat for the Black Death, the culprit was, in fact, the bacterium *Yersinia pestis*. Even the fleas that transmitted the bacteria from the rats to humans fell victim, dying of starvation as the bacteria blocked their stomachs. But the bite or scratch of a rat can spread deadly diseases too.

In 2007, a British woman, Carol Colburn, 56, from Brighton, was scratched by a rat she was trying to free from a bird feeder. Within days she developed flu-like symptoms, which rapidly worsened to a yellowing of the skin and partial paralysis. She was rushed into hospital but died

of a heart attack. Her illness was traced back to the injuries from the rat, which had infected her with Weil's disease, or *Leptospirosis*, a very rare disease that is carried by small rodents.

The threat from rats still exists, although such instances are very rare. The latest World Health Organisation survey shows just five cases of people dying as the result of a rat bite: one in each of Mexico, Colombia, Brazil, South Korea and the USA.

SEA MONSTERS

Old mariners' tales of fearsome sea monsters are part of our folklore, yet the lack of such sightings in the modern era makes it seem unlikely that these Krakens and sea serpents ever really existed. Many of their depictions in maps and drawings appear to be fanciful concepts from the illustrator's imagination, or exaggerations of the lurid descriptions of sailors.

Yet the tales are so numerous, there has to be some basis for them.

In the 19th Century, the Royal Navy kept a record of these sightings. For example, in 1830 the captain of the *Rob Roy* submitted a report to the Admiralty about a beast that had been spotted by his crew near St Helena in the South Atlantic. The report describes the creature as 'a thundering great sea snake', measuring (rather precisely) 'about 129ft in length' (40m).

In 1857, a similar sighting was reported near St Helena, this time by Commander George Harrington of *The Castilian*.

The deep ocean is the last great frontier. We've been further into Space than we have beneath the waves of our own planet. The average depth of the world's oceans is 2.4 miles (3.8km), plunging to 6.8 miles (11km) at its deepest point. That's a lot of water for a large beastie to hide in.

We are still discovering evidence of creatures, alive and dead, that lend credence to the old legends of sea monsters. Archaeologists have found fossils of a

gigantic prehistoric monster that measured 50ft (15m) in length, with a 10ft-long (3m) head and teeth the size of cucumbers. Its jaws had the crushing power of a car compacter.

The creature, believed to be a type of Plesiosaur, lived 150 million years ago. But who is to say it ever became extinct?

There are, in fact, plenty of impossibly ugly fish down there that do quite closely resemble the sea monsters of old. The Grenadier (or Rattail), for example, is a vile-looking fish that has a lizard-like tail and a huge head and mouth that looks not dissimilar to the mouth of the Kraken when viewed up close. It's not a pretty sight.

The concept of the big, gnashy teeth could well have derived from the fangtooth, dragonfish, viperfish, or numerous other deep sea fish whose tooth-to-body size ratio makes the Great White shark look positively tame. And as for those huge mouths? Take a look at the megamouth

shark. This rare shark grows up to 18ft (5.5m) long, with a gaping mouth that looks more than ready to swallow a man whole. In fact, it's designed for swallowing plankton.

The stories of 'thundering great sea snakes' were probably based on sightings of the oarfish. Growing to 38ft (10m) and more, this long, thin fish spends most of its time at depths of more than 650ft (200m) and is usually in distress when spotted near the surface. In this state, it tends to swim with its head out of the water, just like a sea serpent.

So it's very likely that the monsters reported in seamen's tales and depicted on old maritime maps were based in fact, even if the fact was probably a little smaller and less aggressive than the fiction. There are plenty of hideous sea beasts that would no doubt have given a bathing sailor a scare, but they're unlikely to have swallowed his ship.

Of course, that all depends on the size of the ship. Mexican fishermen are all too aware of the aggressive behaviour of the Humboldt squid. These vicious animals grow very quickly up to 6ft (1.8m) long and hunt in packs. They have been known to attack divers in the water. Once they grip hold of something with their suckers, they're almost impossible to prise off. And if they get their powerful beak into you, you're lunch.

The Mexicans call them *diablos rojos* (red devils) because of their bioluminescence – the ability to change colour. They flash red when

hunting, probably to mesmerize their prey. There are tales of Mexican fishermen being dragged from their boat and chewed up beyond recognition.

If a 6ft (1.8m) Humboldt squid will do this, imagine what a 45-footer (14m) might do. That's the estimated size to which a colossal squid grows. These monsters of the deep were only discovered in 1925, when a couple of tentacles were found in the stomach of a sperm whale. Only a handful of complete specimens have been landed, the largest of which was reckoned to be 33ft (10.5m) in total length, though it had shrunk to about 14ft (4m) by the time it was properly examined.

By comparing the dimensions of this and other specimens to body parts

found in other whales and sharks, scientists have deduced that somewhere out there, in the depths of the ocean, there exist colossal squid the size of a small fishing trawler, with eyes 20in (51cm) across.

It's enough to make you think twice about ordering the calamari.

A HIGHER POWER

What are the odds?

1 in 500 – being abducted by aliens

1 in 10,000 – being possessed by the devil

1 in 3,500 – being struck by lightning

1 in 1,700 – being killed by the weather

ACTS OF GOD

Sometimes Mother Earth can be a vengeful hostess, unleashing her power with a deadly ruthlessness that, despite the best efforts of modern science, we can't always predict. Did you know that in America, for example, your chances of being killed by extreme weather conditions are **1 in 76,000** each year?

If there is a sense that deadly weather extremes are becoming more frequent, the figures give comfort.

On a global scale, your risk of being killed by extreme weather is only 1/20th of what it was a hundred years ago. The odds of the weather being the cause of your death are **1 in 1,700**.

And the annual risk of being killed by the weather is **3 in 1,000,000**.

The biggest threat a hundred years ago was drought. For most of the last century, the mortality rate was around 60 per million people per year. But relief operations have reduced that considerably to **0.03 per million** in the last 20 years.

The mortal threat from flooding has also been hugely reduced, from **32 per million to 1.3**.

But the element we can't tame is wind. Windstorms have gone from being the third-biggest elemental threat to being the biggest, the annual risk of **2.45 deaths per million** having fallen very little from the **4 per million** of the 20th century.

STORMY WEATHER

Anyone who's watched *The Wizard of Oz* will be well aware of the risks of being carried away by a tornado in Kansas. The USA is hit by more twisters than any other country on earth – 1,200 a year on average – and they occur most frequently in Kansas and Oklahoma, around the month of May.

It's not the sort of thing you expect to see in England's Shakespeare country, yet the UK experiences more tornadoes than any other European country – around 30 per year.

Blowing in the wind

Admittedly, most of them are barely strong enough to remove a lady's hat, but occasionally the mild-weathered UK is forced to endure the ravages of nature.

In 2005, an area of Birmingham, just 30 miles from Stratford-upon-Avon, was blasted by a tornado measuring T4 on the Torro scale (the scale ranges from T1 up to T10).

Winds of up to 136mph (219km/h) wrecked a supermarket, tore up trees and damaged more than 100 homes, causing injuries to 19 people.

Eye-witness Jane Trobridge reported, 'There were a couple of really loud thunder claps and bright lightning strikes and then the wind picked up. We looked out of the window and there was loads of debris flying around – bits of trees, roof insulation and rubbish.

'Suddenly everything started blowing upwards. I tried to get out to get a better look, but I couldn't force the door open. I could see loads of flying debris circling, and then the roof seemed to lift off the petrol station and swing almost completely upright and then fall back down.'

The tornado lasted just four minutes but did enough damage in that short time to rank as the worst to hit Britain since 1931. Worryingly for residents, that previous tornado hit the same area of Birmingham, killing one person.

WHEN THE EARTH MOVES

The San Francisco earthquake of 1906 is the worst natural disaster in American history, claiming an estimated 3,000-plus lives, yet in terms of casualties it falls way down the list of global disasters. The earthquake that devastated Haiti in 2010 was less powerful than the San Francisco quake, yet it killed 222,500.

Earthquakes occur with frightening regularity and in locations all around the world. On average there are 16 earthquakes of Force 7 magnitude or more each year – 7 being the level at which they are deemed to be major earthquakes. Many of these occur well away from human civilization, but when they do hit heavily populated areas the effect is devastating, especially if the people are living in constructions not designed to withstand major earthquakes.

The deadliest earthquake on record claimed 820,000 lives in and around Shaanxi, China, in 1556. Most of the population of the region lived in caves dug out of the earth. Sixty per cent lost their lives.

TSUNAMI

Up until 2004, people embarking on exotic holidays didn't tend to consider the likelihood of being struck by a natural disaster. That was the furthest thing from their mind as they reclined on a beach in South-east Asia, sipping a cocktail and gazing wistfully at the horizon.

The tsunami that crashed in on 26 December changed all that.

The word 'tsunami' was new to many of us. The more common expression had been 'tidal wave', a misnomer, since these massive waves are not tidal. They are formed by massive displacements of water, usually brought about by seismic shifts (earthquakes) or volcanic eruptions. In deep sea this displacement of water spreads at around 600mph (966km/h), though its effect on the surface is barely noticeable. It's when the wave reaches shallow water and slows down that it builds in height.

The earthquake that caused the 2004 tsunami measured 9.1 in magnitude and caused waves as high as 98ft (30m) in some places. It travelled 5,000 miles (8,000km) across the Indian Ocean to wreak havoc in Somalia, Tanzania and even South Africa.

The fact that the 2004 tsunami was so massive and struck without

warning, causing so much devastation in an area not normally associated with the phenomenon, marked it out as a freak occurrence.

Yet tsunamis happen about five times a year on average. Fortunately most of them occur in sparsely populated areas of the Pacific Ocean and are minuscule by comparison.

As a result of the 2004 tsunami, the Indian Ocean is now recognized as a major danger area and early warning systems are being put in place. An estimated 280,000 people died in that disaster.

Next time something similar happens, the chances of survival should be considerably higher.

A HARD RAIN'S GONNA FALL

Hailstorms can sting a bit if you get caught out in one, but you don't usually expect them to threaten life and limb. Most hailstones measure about a quarter of an inch in diameter and look quite pretty as they bounce down off the roof. But

sometimes they don't bounce off – they go straight through it.

In July 2010 Leslie Scott of Vivian, South Dakota found a hailstone that measured 8in (20.5cm) in diameter and weighed just under 2lb (1kg). It had crashed through his roof, leaving a gaping hole. Scott claimed the hailstone had shrunk by about 2in (5cm) by the time it was officially measured because the power to his freezer had been knocked out by the storm.

Miraculously, the hailstorm, which smashed car windows and left deep dents in the ground, didn't cause any serious injuries. However, in Bangladesh in 1986 local citizens were not so lucky. More than 90 people were killed when hailstones the size of grapefruits bombarded them from the sky.

THERE'S GONNA BE AN ERUPTION!

Every year, 2.6 million tourists wander among the ruins of Pompeii and Herculaneum, treading in the shadow of Mt Vesuvius, which buried

these Roman towns in volcanic ash in AD79. They marvel at the preserved remains, the graffiti, the chariot tracks. They seem oddly oblivious to the fact that they are walking on an explosive device of epic proportions.

Vesuvius has not finished erupting. Since AD79 it has erupted 35 times, an average of once every 55 years. The last time was in 1944, a relatively minor eruption that nevertheless left 26 dead. That's nothing on the thousands who were wiped out in AD79. By the law of averages, the next eruption is overdue, and the longer it waits, the greater the chance that Vesuvius will blow its top in catastrophic style. Some experts believe there is a high probability of an explosive eruption of a similar magnitude to the one that buried Pompeii by the end of this century.

KILLER LAKES

If you're travelling in Cameroon, beware of the exploding lakes. In 1984, villagers living near Lake Monoun were suddenly asphyxiated by a cloud of gas, which descended on them out of nowhere. It was a slient and terrible tragedy. Thirty-seven people died. Two years later, a similar gas outbreak near Lake Nyos had an even more devastating effect, killing 1,700 people and their livestock.

Lake Nyos and Lake Monoun lie in volcanic craters, where carbon dioxide leaks out of the rock, turning the water to acid. When a limnic eruption occurs, the carbon dioxide is released from the water in huge volumes, moving silently and suffocating everything in its path.

A BOLT FROM THE BLUE

Lightning never strikes twice in the same place, so the saying goes. But it has been known to strike more than once in the vicinity of the same person. Roy Sullivan, a ranger at Shenandoah National Park in Virginia, USA, was struck by lightning seven times in 35 years, between 1942 and 1977. All seven were verified by doctors and officially recorded by the superintendant of Shenandoah National Park. But Sullivan claimed he had also been struck when he was a child, this one going unrecorded.

The fact that he worked outdoors in an area with a high incidence of thunderstorms made Sullivan a likely target for lightning strikes, a phenomenon that kills around 50 people a year in the USA alone, but

Pos	Top 10 Countries for fatal lightning Strikes		
Pos	Country	Deaths	Odds
1	Mexico	223 deaths	1/500,000
2	Thailand	171 deaths	1/395,000
3	South Africa	150 deaths	1/320,000
4	Brazil	132 deaths	1/1,450,000
5	Romania	75 deaths	1/285,000
6	Colombia	71 deaths	1/640,000
7	Cuba	70 deaths	1/160,000
8	Peru	68 deaths	1/420,000
9	United States	50 deaths	1/6,200,000
10	Panama	17 deaths	1/200,000

his account of the fifth and sixth strikes suggest that he was being singled out. On both occasions he saw a storm cloud forming and did his utmost to drive away from it, but the cloud caught up with him and he was struck despite his best efforts.

The odds of being struck once in a lifetime are around **1 in 3,500**, taking 70 years as an average lifetime. That makes Sullivan's odds $1/3,500^8$ if we count his first strike when he was a child. Or, written another way, **1 in 2,250,000,000,000,000,000,000,000,000,000**. The tragic postscript to Roy Sullivan's amazing story is that he took his own life at the age of 71, by shooting himself in the stomach.

According to the World Health Organisation, around 1,200 people worldwide are killed each year by lightning, with Mexico suffering the most fatalities. However, the place with the most deadly odds is Cuba, where **1 in 160,000** people are killed by lightning strikes each year.

WHEN YOUR LUCK'S IN

Retired Croatian music teacher Frane Selak considers himself the luckiest man in the world, and with good reason. In 2003 he won the equivalent of US$1 million on the Croatian lottery with his first ticket, but that is a mere footnote in a story of incredible escapes from near-death experiences.

In 1962, Selak managed to survive when a train he was travelling in left the track and plunged into an icy river, killing 17 passengers. The following year, 19 died when a plane travelling inside Yugoslavia crashed, but Selak landed in a haystack, after falling 2,780ft (850m), and lived to tell the tale. Five years on, he again found himself in a freezing river when a bus he was in crashed off a bridge. Four people died, but he and the driver survived.

Selak then escaped on two separate occasions when his car caught fire, and finally he survived when his car was forced off the road by a UN armoured truck. He was thrown clear into a tree as his car plunged 500ft (150m) and blew up.

You could say he was lucky to escape all that AND win the lottery. But having been unlucky enough to have to endure so many traumas, wasn't a win on the lottery the least he deserved?

What are the odds?	
Escaping a fatal train crash	1/20
Escaping a fatal plane crash	1/1,000
Escaping a fatal bus crash	1/32
Escaping an exploding car	1/2
Winning the lottery once in a lifetime	1/3,840
Frane Selak's cumulative odds	1/4,900,000

CLOSE ENCOUNTERS

'The chances against anything manlike on Mars are a million to one,' says the astronomer Ogilvy in HG Wells' sci-fi classic *The War of the Worlds*. And while subsequent NASA probes appear to have proven fairly conclusively that there are no little green men populating the red planet, the likelihood of life as we know it existing elsewhere in the Universe is strong, perhaps even infinitesimally so, depending on which theory you choose to believe.

According to some people – a surprisingly large number, in fact – the extra-terrestrials have already started visiting Earth. And they've been taking away souvenirs… human ones!

Alien abduction isn't just a convenient excuse for being late for work or forgetting to do your homework.

Ever since Betty and Barney Hill made the world sit up and take notice in 1961 with their story of a close encounter with a UFO following them down Route 3 through New Hampshire, vast numbers of people all over the world reckon they have been abducted by aliens, and not just the once either.

1 in 500: that's the remarkable number of Americans who have experienced alien abduction, according to a 1998 poll.

First commissioned in 1991 by the National Institute of Discovery Sciences (NIDS), the 1998 Roper Survey into Unusual Personal Experiences quizzed 6,000 respondents on their experiences of phenomena that were taken to be indicators of alien abduction. A respondent was classified as a possible abductee if they had experienced four of the following five key indicators more than once:-

1. Waking paralyzed and sensing a presence in the room
2. Missing time
3. Feeling of flying
4. Seeing balls of light
5. Anomalous scars

0.2 per cent of people polled claimed to have experienced all five!

It is generally accepted amongst UFOlogists that alien abduction isn't something that only happens to Americans on deserted country roads in Nebraska. In fact, it is considered to be evenly spread around the world, meaning that, if we are to accept the NIDS survey, 0.2 per cent of the world's population has been abducted by aliens more than once in their life.

That's more than 13 million people!

To put that into perspective, of the 84,500 people who attended the 2010 soccer World Cup Final in Johannesburg, 169 are likely to have been abducted by aliens.

However, few of them have produced the quality of evidence to match the Hills' gripping account. Having undergone hypnotherapy, Betty and Barney produced closely correlating descriptions of their ordeal, culminating in Betty drawing a detailed map of the Zeta Reticuli star system, six years before human astronomers discovered it!

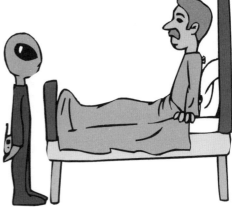

WHAT'S THE DANGER OF BEING STRUCK BY A METEORITE?

Of course, alien spacecraft aren't the only extra-
terrestrials that pose a threat to mankind.
Scientists estimate that anything between
18,000 and 84,000 meteorites
weighing 4oz (10g) or more crash
into the earth every year. That's
between 36 and 166 per million
sq km (386,000 sq miles).

They generally range in size from a small pebble to a basketball, but
they can come considerably bigger than that. The Morokweng Crater
in the Kalahari Desert in South Africa, discovered in 1994, was made
by an asteroid measuring around five miles (8km) in diameter.

And a similar-sized lump of space rock is thought to have caused the
Tunguska Event in Siberia in 1902, when an explosion estimated at
1,000 times the magnitude of the Hiroshima atom bomb flattened trees
over an area of more than 750 sq miles (2,000 sq km) and sent out a
shock wave that knocked people off their feet hundreds of miles away.

Both of these strikes would have been sufficient to destroy a city,
yet fortunately these rocks have a habit of falling in the most sparsely
populated areas of the planet and nobody seems to get hit by them.
Having said that, had there been anyone standing under the Tunguska

meteoroid, they wouldn't have been able to tell anyone about it.

The first meteorite on record to have injured a human appeared as recently as 1954. On November 30th, 31-year-old Ann Elizabeth Hodges, of Sylacauga, Alabama, was enjoying a siesta on the sofa after lunch when a piece of rock the size of a shotputt crashed through the roof, bounced off the radio and hit her in the side, waking her up rather abruptly.

Hodges suffered severe bruising and a storm of international media interest. After fighting off her landlord and the US military, both of whom wanted custody of the meteorite, she donated it to the Alabama Museum of Natural History, where it can be seen today.

Hodges, who died in 1972, remains one of only a handful of people who have been hit by meteorites. And there are no confirmed cases of anyone being killed by one.

So you can understand Bosnian Radivoje Lajic's paranoia when a sixth meteorite in three years landed on his house in July 2010. 'I have no doubt I am being targeted by aliens,' said Lajic.

'The chance of being hit by a meteorite is so small that getting hit six times has to be deliberate.'

GETTING CLOSE TO A GHOST

There are two kinds of people in the world: those who believe in ghosts and those who don't. The latter sometimes switch sides, usually as a result of some paranormal experience. Many of the former never see proof of what they believe in, and probably wouldn't want to. Suffice it to say that the odds of believing in ghosts are considerably higher than the odds of seeing one. But the more you believe in them, the more likely you are to see them. Draw your own conclusions.

Most of us go through life hoping not to see a ghost – when we turn off the light and climb the stairs to bed; when we go down into the cellar to find candles during a power failure; when we find ourselves on

holiday by mistake in a rickety old house in the middle of nowhere, during a thunderstorm…

But there are places in the world where it is not uncommon to actually marry a ghost. Yes, that's right. Ghost marriage is a longstanding tradition in China, whereby the bride or groom, or sometimes both, are deceased. The idea is to provide the unmarried spirit with a partner in the afterlife.

The practice can also be found in other cultures in which

spiritualism plays a cental part, such as Sudan and India. But it is also, you may be surprised to learn, a legal process in France.

Ever since 1959, when President de Gaulle granted a widow the right to marry her fiancé, who had been drowned in the Fréjus dam disaster, French law has permitted civilian marriages to deceased partners, subject to approval from the President.

DEMONIC POSSESSION

One man who does not believe in ghosts is Father Amorth, the Vatican's chief exorcist. Ghosts, he says, are a figment of the human mind. The Devil, on the other hand, is very real, according to Amorth, and more and more people seem to be getting possessed by him.

Amorth claims to have carried out over 70,000 exorcisms in a 60-year career. And it can be a messy process. Father Amorth has witnessed people vomiting needles and broken glass, often over him. Projectile vomiting is a speciality of the Devil, apparently. That 70,000 equates to far fewer actual victims, as most people have to undergo more than one exorcism, sometimes hundreds even, before the demons are driven out completely. Yet the number of people seeking the services of an exorcist has grown exponentially in recent years, and the number of exorcists in the world has risen accordingly. The chances of being possessed by the Devil – or thinking you are – are still very small, however. Even in Italy, where there are around 300 exorcists, the number of people seeking their services amounts to fewer than **1 in 10,000.**

sources

2009 Homeless Assessment Report to
Congress

American Heart Association

American Society for Aesthetic Plastic
Surgery

Cancer Research

Cantu et al. Neurosurgery 52:846-853,
2003. Brain injury-related fatalities in
American football, 1945–1999

Catalano S. Intimate Partner Violence
in the United States. Bureau of Justice
Statistics. December 19, 2007

Centers for Disease Control and
Prevention in Atlanta

Centers for Disease Control and
Prevention. Youth Risk Behavior
Surveillance, United States 2009.
Surveillance Summaries, June 10,
2010. MMWR 2010;59(No. SS-5)

CIA World Factbook

Clifton, Dog Attack Deaths and
Maimings, US & Canada, September
1982 to November 13, 2006.

Dropzone.com

Eaton, D. et al. Youth Risk Behavior
Surveillance — United States, 2009.
Department Of Health And Human
Services Centers for Disease Control
and Prevention. June 4, 2010, Vol. 59.
No. SS-5

FlightStats

Food Allergy and Anaphylaxis Network
(FAAN)

Forbes

Gender Variance in the UK: Prevalence,
Incidence, Growth and Geographic
Distribution

Health and Safety Executive: Risk
Education Statistics, http://www.hse.
gov.uk/education/statistics.htm#death

Holeinone.com

International Action Network on Small
Arms (IANSA),

National Geographic Channel

National Safety Council

National Vital Statistics System

New York State Department of Health

Number of jumps made in 2006 from
2006 membership survey results,
http://www.uspa.org/about/images/
memsurvey06.pdf

OAG Aviation

PlaneCrashInfo.com accident database

PlaneCrashInfo.com accident database
1985–2009

Redelmeier et al. BMJ
2007;335;1275-1277. Competing
risks of mortality with marathons:
retrospective analysis

Royal Astronomical Society

Royal Society for the Prevention of
Accidents

Seedmagazine.com

Soreide et al. J Trauma.
2007;62:1113-1117. How Dangerous
is BASE Jumping? An Analysis of
Adverse Events in 20,850 Jumps From
the Kjerag Massif, Norway

St. George, D. 'Textual harassment' a
weapon in dating violence. Richmond
Times-Dispatch, July 24, 2010.

Teen Dating Abuse 2009 Key Topline
Findings. Family Violence Prevention
Fund. June 10, 2009.

The 2009 Annual Homeless Assessment
Report to Congress. US Department
of Housing and Urban Development.

The Dart Book by Kari Kaitanen. ISBN
951-96680-1-2

The International Shark Attack File

The Joint Commission

The New England Journal of Medicine

The Skin Cancer Foundation

Turk et al. Br. J. Sports Med.
2008;42;604-608. Natural and
traumatic sports-related fatalities: a
10-year retrospective study

US Centers for Disease Control and
Prevention

UNAIDS 2009 AIDS Epidemic Update

UNdata – data.un.org

United Nations Office on Drugs and
Crime

United States Parachute Association
accident statistics, http://www.uspa.
org/about/page2/relative_safety.htm

US Census Bureau

Westman et al. Accident Analysis and
Prevention 37 (2005) 1040-1048.
Fatalities in Swedish Skydiving

Which?

World Bank, World Development
Indicators

*World Christian Encyclopedia: A
comparative survey of churches and
religions – AD30 to 2200.* Oxford
University Press, (2001)

World Health Organisation

World Heart Federation

World Institute for Development
Economics Research at United Nations
University

www.adventurestats.com

www.expatriate-solutions.net